T0291989

LONDON MATHEMATICAL SOCIETY LECTURE NOTE SERIES

Managing Editor: Professor J.W.S. Cassels, Department of Pure Mathematics and Mathematical Statistics, 16 Mill Lane, Cambridge CB2 1SB, England

4 Algebraic topology, J.F.ADAMS
5 Commutative algebra, J.T.KNIGHT
8 Integration and harmonic analysis on compact groups, R.E.EDWARDS
11 New developments in topology, G.SEGAL (ed)
12 Symposium on complex analysis, J.CLUNIE & W.K.HAYMAN (eds)
13 Combinatorics, T.P.McDONOUGH & V.C.MAVRON (eds)
15 An introduction to topological groups, P.J.HIGGINS
16 Topics in finite groups, T.M.GAGEN
17 Differential germs and catastrophes, Th.BROCKER & L.LANDER
18 A geometric approach to homology theory, S.BUONCRISTIANO, C.P.ROURKE & B.J.SANDERSON
20 Sheaf theory, B.R.TENNISON
21 Automatic continuity of linear operators, A.M.SINCLAIR
23 Parallelisms of complete designs, P.J.CAMERON
24 The topology of Stiefel manifolds, I.M.JAMES
25 Lie groups and compact groups, J.F.PRICE
26 Transformation groups, C.KOSNIOWSKI (ed)
27 Skew field constructions, P.M.COHN
29 Pontryagin duality and the structure of LCA groups, S.A.MORRIS
30 Interaction models, N.L.BIGGS
31 Continuous crossed products and type III von Neumann algebras,A.VAN DAELE
32 Uniform algebras and Jensen measures, T.W.GAMELIN
34 Representation theory of Lie groups, M.F. ATIYAH et al.
35 Trace ideals and their applications, B.SIMON
36 Homological group theory, C.T.C.WALL (ed)
37 Partially ordered rings and semi-algebraic geometry, G.W.BRUMFIEL
38 Surveys in combinatorics, B.BOLLOBAS (ed)
39 Affine sets and affine groups, D.G.NORTHCOTT
40 Introduction to Hp spaces, P.J.KOOSIS
41 Theory and applications of Hopf bifurcation, B.D.HASSARD, N.D.KAZARINOFF & Y-H.WAN
42 Topics in the theory of group presentations, D.L.JOHNSON
43 Graphs, codes and designs, P.J.CAMERON & J.H.VAN LINT
44 Z/2-homotopy theory, M.C.CRABB
45 Recursion theory: its generalisations and applications, F.R.DRAKE & S.S.WAINER (eds)
46 p-adic analysis: a short course on recent work, N.KOBLITZ
47 Coding the Universe, A.BELLER, R.JENSEN & P.WELCH
48 Low-dimensional topology, R.BROWN & T.L.THICKSTUN (eds)
49 Finite geometries and designs,P.CAMERON, J.W.P.HIRSCHFELD & D.R.HUGHES (eds)
50 Commutator calculus and groups of homotopy classes, H.J.BAUES
51 Synthetic differential geometry, A.KOCK
52 Combinatorics, H.N.V.TEMPERLEY (ed)
53 Singularity theory, V.I.ARNOLD
54 Markov process and related problems of analysis, E.B.DYNKIN
55 Ordered permutation groups, A.M.W.GLASS
56 Journêes arithmêtiques, J.V.ARMITAGE (ed)
57 Techniques of geometric topology, R.A.FENN
58 Singularities of smooth functions and maps, J.A.MARTINET
59 Applicable differential geometry, M.CRAMPIN & F.A.E.PIRANI
60 Integrable systems, S.P.NOVIKOV et al
61 The core model, A.DODD

62 Economics for mathematicians, J.W.S.CASSELS
63 Continuous semigroups in Banach algebras, A.M.SINCLAIR
64 Basic concepts of enriched category theory, G.M.KELLY
65 Several complex variables and complex manifolds I, M.J.FIELD
66 Several complex variables and complex manifolds II, M.J.FIELD
67 Classification problems in ergodic theory, W.PARRY & S.TUNCEL
68 Complex algebraic surfaces, A.BEAUVILLE
69 Representation theory, I.M.GELFAND et al.
70 Stochastic differential equations on manifolds, K.D.ELWORTHY
71 Groups - St Andrews 1981, C.M.CAMPBELL & E.F.ROBERTSON (eds)
72 Commutative algebra: Durham 1981, R.Y.SHARP (ed)
73 Riemann surfaces: a view towards several complex variables, A.T.HUCKLEBERRY
74 Symmetric designs: an algebraic approach, E.S.LANDER
75 New geometric splittings of classical knots, L.SIEBENMANN & F.BONAHON
76 Linear differential operators, H.O.CORDES
77 Isolated singular points on complete intersections, E.J.N.LOOIJENGA
78 A primer on Riemann surfaces, A.F.BEARDON
79 Probability, statistics and analysis, J.F.C.KINGMAN & G.E.H.REUTER (eds)
80 Introduction to the representation theory of compact and locally compact groups, A.ROBERT
81 Skew fields, P.K.DRAXL
82 Surveys in combinatorics, E.K.LLOYD (ed)
83 Homogeneous structures on Riemannian manifolds, F.TRICERRI & L.VANHECKE
84 Finite group algebras and their modules, P.LANDROCK
85 Solitons, P.G.DRAZIN
86 Topological topics, I.M.JAMES (ed)
87 Surveys in set theory, A.R.D.MATHIAS (ed)
88 FPF ring theory, C.FAITH & S.PAGE
89 An F-space sampler, N.J.KALTON, N.T.PECK & J.W.ROBERTS
90 Polytopes and symmetry, S.A.ROBERTSON
91 Classgroups of group rings, M.J.TAYLOR
92 Representation of rings over skew fields, A.H.SCHOFIELD
93 Aspects of topology, I.M.JAMES & E.H.KRONHEIMER (eds)
94 Representations of general linear groups, G.D.JAMES
95 Low-dimensional topology 1982, R.A.FENN (ed)
96 Diophantine equations over function fields, R.C.MASON
97 Varieties of constructive mathematics, D.S.BRIDGES & F.RICHMAN
98 Localization in Noetherian rings, A.V.JATEGAONKAR
99 Methods of differential geometry in algebraic topology, M.KAROUBI & C.LERUSTE
100 Stopping time techniques for analysts and probabilists, L.EGGHE
101 Groups and geometry, ROGER C.LYNDON
102 Topology of the automorphism group of a free group, S.M.GERSTEN
103 Surveys in combinatorics 1985, I.ANDERSEN (ed)
104 Elliptical structures on 3-manifolds, C.B.THOMAS
105 A local spectral theory for closed operators, I.ERDELYI & WANG SHENGWANG
106 Syzygies, E.G.EVANS & P.GRIFFITH

London Mathematical Society Lecture Note Series. 105

A Local Spectral Theory for Closed Operators

IVAN ERDELYI
Temple University
WANG SHENGWANG
Nanjing University

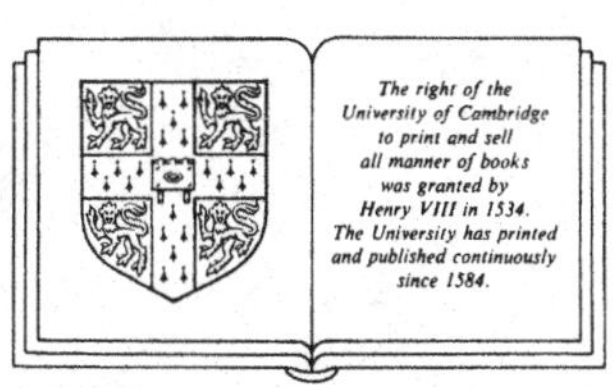

CAMBRIDGE UNIVERSITY PRESS
Cambridge
London New York New Rochelle
Melbourne Sydney

Published by the Press Syndicate of the University of Cambridge
The Pitt Building, Trumpington Street, Cambridge CB2 1RP
32 East 57th Street, New York, NY 10022, USA
10, Stamford Road, Oakleigh, Melbourne 3166, Australia

© Cambridge University Press 1985

First published 1985

Library of Congress catalogue card number: 85-47812

British Library cataloguing in publication data

Erdelyi, I.
A local spectral theory for closed operators- (London Mathematical Society lecture note series, ISSN 0076-0052; 105)
1. Banach spaces 2. Spectral theory (mathematics)
I. Title II. Wang, Shengwang III. Series
515.7'32 QA322.2

ISBN 0 521 31314 7

Transferred to digital printing 2001

CONTENTS

PREFACE

This monograph has grown out of a recent sequel of jointly written papers by the authors. It proposes to present a unified treatment of the local spectral theory for closed operators acting on a complex Banach space.

While working with closed operators, in a few instances it will be unavoidable to transgress the former concept. More to the point, the theory makes frequent use of operators *coinduced* by closed operators on a quotient space. Such operators, in general, are not even closable. Special efforts will be requested to exhibit conditions which make them, at least, closable and hence useful instruments in the subsequent theory.

The plan of this work may be sketched as follows. After a brief presentation of the *spectral decomposition problem*, Chapter I introduces the notion of the *single valued extension property* with some of its implications. Subsequently, a general study of *invariant subspaces* is followed by the developing ideas for some special types of subspaces, such as ν-*spaces*, μ-*spaces*, *analytically invariant subspaces*, T-*absorbent spaces*, *spectral maximal* and T-*bounded spectral maximal spaces*, T being the given closed operator.

With Chapter II we come to the essence of the spectral theory: the general type of *spectral decomposition property* with its relationship to the *unbounded decomposable operators*.

Chapter III is devoted to the *spectral duality* theory. After having overcome some difficulties due to the general type of the spectral decomposition problem, it was gratifying to have obtained a spectral duality theorem, under the natural constraints of the problem.

Chapter IV provides more structure to the spectral decomposition by means of the *spectral resolvent* concept. The treatment of the somewhat streamlined construct of *spectral capacity* concludes the chapter.

Up to this point, the sections follow an essentially prescribed order. In the last Chapter V, the sections are apparently, independent. They have, however, a unifying theme: *strong decomposability*.

While our notation and terminology are standard, the great variety of definitions, operators and subspaces derived from the basic ones, have forced us to introduce new symbols. A *glossary of notations and symbols* is provided and it should be consulted by the reader.

The definitions, lemmas, propositions, theorems, corollaries, remarks and examples that make up a section are numbered in the order of their appearance. When citing, they will be referred to by the corresponding numbers. The numbered equations, relations, etc., will be quoted by their numbers enclosed by parentheses.

There is a considerable number of excellent research papers on the local spectral theory that has not been included in the bibliography. The bibliography consists of works which are either directly referred to (indicated by an asterisk), or related to the theory developed in this volume.

When it comes the time to express our thanks to those who encouraged us to write this volume, and there is no shortage of them, we cannot help but evoke a comment by the referee of one of our joint papers: "...this paper and [·] are the only papers I know where a human being has gone to the third adjoint (!) of a Banach space and come back with interesting information." We recognize in these words the imaginative and kind thoughts of a great analyst of our time: Robert G. Bartle. But the referee of a paper is anonymus, our allusion is the only conjecture in this volume. Robert G. Bartle has showed much interest in our work, his inspiration and stimulation have helped to pursue our goals. This work is gratefully dedicated to him.

It is our pleasant privelege to thank Professors Cheo, Peh-Hsiun and Tseng, Yuan-Yong from Nanjing University, Xia, Dao Xing (Fudan University, Shanghai) and Jiang, Ze-Jian (Jilin University, Chang Chun) for their help and encouragement.

We gratefully acknowledge the support of both the Chinese Academy of Science and Temple University for granting one of the authors research funds and the other a Summer Research Fellowship in 1983, respectively. The generosity of the above named institutions allowed the authors an undisturbed period of time for research related to the present volume.

Finally, we wish to express our deepest gratitude to E. Brian Davies and Rufus Neal, editors of the London Mathematical Society publications and Cambridge University Press for their kind and helpful cooperation at every stage in the production of this volume.

Ivan Erdelyi and Wang, Shengwang (王 声 望)

GLOSSARY OF NOTATIONS AND SYMBOLS

$AI(T)$: the set of analytically invariant subspaces under T, 18.

A_T : the class of functions, analytic on a neighborhood of $\sigma_\infty(T)$, 4.

$B(X)$: the Banach algebra of bounded linear operators $T : X \to X$, 4.

$\mathbb{C}$: the complex field.

$\mathbb{C}_\infty = \mathbb{C} \cup \{\infty\}$: the compactified $\mathbb{C}$.

S^c : the complement of a set S.

$C[0,1]$: the space of continuous functions on $[0,1]$.

cov S : the collection of all finite open covers of a set $S \subset \mathbb{C}$.

$\mathcal{D}_T$: the domain of an operator T.

Δ : Cauchy domain, 3.

∂S : the boundary of a set S.

(*), (**), (***), (****) : domain-density conditions, 74.

E : spectral resolvent, 81.

$\tilde{E}$: prespectral resolvent, 84.

η : (pre) spectral capacity, 102.

η': 103.

F and F^K : the family of all closed and compact subsets of $\mathbb{C}$, respectively, 81.

G and G^K : the family of all open and relatively compact open subsets of $\mathbb{C}$, resp.

G_F : 82.

$G(\cdot)$: the graph of an operator, 12.

$VG(\cdot)$: the inverse graph of an operator, 13.

H Inv T : the lattice of all hyperinvariant subspaces under T, 129.

I : the identity operator

$\hat{I}_\alpha$: the identity in $\hat{X}_\alpha$, 138.

Int S : the interior of a set S.

Inv T : the lattice of all invariant subspaces under T.

$J : X \to X^{**}$: the canonical embedding of X into X^{**}, 73.

$K : X^* \to X^{***}$: the canonical embedding of X^* into X^{***}, 73.

$N(\cdot)$: null space.

$\mathbb{N}$: the set of natural numbers.

$Z^+ = \mathbb{N} \cup \{0\}$.

ω_f, $\omega_{\hat{f}}$: domain of f, $\hat{f}$.

P (as in §.10 and Appendix A): the projection of X^{***} onto KX^* along $(JX)^a$, 74.

$\mathbb{R}$: the field of real numbers.

R_T or $R(T)$: the range of an operator T.

$R(\cdot;T)$: the resolvent operator.

$\rho(T)$: the resolvent set of T.

$\rho_\infty(T)$: the unbounded component of $\rho(T)$, if $T \in B(X)$.

$\rho(x,T)$: the local resolvent set of T at $x \in X$.

$S(\cdot)$: lattice of subspaces, 81.

(n-) SDI : the (n-) spectral decomposition property with respect to the identity, 129.

(n-) SDP : the (n-) spectral decomposition property, 33.

SM(T) : the set of all spectral maximal spaces of T, 24.

$SM_b(T)$: the set of all T-bounded spectral maximal spaces, 24.

SSDP : the strong spectral decomposition property, 108.

Supp : support (of $x \in C[0,1]$, 65).

SVEP : the single valued extension property, 5.

$\sigma(T)$: spectrum of T.

$\sigma_p(T)$: the point spectrum of T.

$\sigma_a(T)$: the approximate point spectrum of T.

$\sigma_\infty(T) = \sigma(T) \cup \{\infty\}$: the extended spectrum of T.

$\sigma(x,T)$ = the local spectrum of T at x.

$T : \mathcal{D}_T(\subset X) \to X$: given closed (if not specified otherwise) operator.

$T|Y$: the restriction of T to Y.

T/Y : the coinduced operator by T on the quotient space X/Y.

U_1, U_2 : complementary simple sets, 63.

V_∞ : neighborhood of ∞, 4.

$V : X \times X \to X \times X$: the mapping $V(x,y) = (-y,x)$, 13.

$\vee$: span.

X : Banach space over $\mathbb{C}$.

$X^*, X^{**}, X^{***}, \ldots$: successive dual spaces; $T^*, T^{**}, T^{***}, \ldots$: successive conjugates of T.

$X(T,\cdot)$: spectral manifold in X, 9.

$\Xi(T,F)$, for $F \subset \mathbb{C}$ compact : T-bounded spectral maximal space associated to $X(T,F)$, 30.

$\Xi(T,G)$, for open $G \subset \mathbb{C}$, 71.

X/Y : quotient space.

$\hat{x}$, $\tilde{x}$, $\mathring{x}$: cosets in various quotient spaces.

$X_{\alpha-0}$: 138.

Y^a : the annihilator of $Y \subset X$ in X^*.

${}^aY$: the preannihilator of $Y \subset X^*$ in X.

Y^α : 150.

Υ : 14.

$\Upsilon(T|Y,\cdot)$: $T|Y$-bounded spectral maximal space associated to $Y(T|Y,\cdot)$, 109.

$\zeta(T|Z,\cdot)$: $T|Z$-bounded spectral maximal space associated to $Z(T|Z,\cdot)$, 112.

$\alpha(x,\lambda,\cdot)$: elementary element, 63.

$\tau(A,B)$: for dual spaces A and B is the topology on A induced by B, 148.

$\cong$: equivalence of subspaces under isometric isomorphism; and unitary equivalence of operators.

$\overline{}^{w}$: weak*-closure.

CHAPTER I. INTRODUCTION

§.1. THE SPECTRAL DECOMPOSITION PROBLEM.

In extending the spectral theory beyond the class of normal operators, N. Dunford [Du.1958] gives a formal definition of the *spectral decomposition* (reduction) problem, for a closed operator T acting on a complex Banach space X. To interpret Dunford's definition, one has to express X as a finite direct sum of invariant subspaces X_i, such that the spectra of the restrictions $T|X_i$ be contained in some given closed sets.

For the class of *spectral operators*, the spectral decomposition is accomplished with the help of a *spectral measure* E: a homomorphic map of the σ-algebra of the Borel sets of the complex plane $\mathbb{C}$ into the Boolean algebra of the projection operators on X, with unit $I = E(\mathbb{C})$. A spectral measure, countably additive in the strong operator topology, is uniquely determined by T (for unbounded spectral operators, see e.g. [Ba.1954]), and is referred to as the *resolution of the identity* (or spectral resolution) of T.

A generalization of the spectral operator concept is due to C. Foiaş [Fo.1960], who replaced the role of the spectral measure by that of the *spectral distribution* (for a full dress account of this theory see [C.1968] and [C-Fo.1968]).

By making the spectral theory independent of such external constraints as direct sum decomposition, spectral measure and spectral distribution, we adopt the following

1.1. DEFINITION. Given a closed operator $T : \mathcal{D}_T(\subset X) \to X$, a *spectral decomposition* of X by T is a finite system

$$\{(G_i, X_i)\} \subset \mathcal{G} \times \text{Inv } T$$

satisfying the following conditions:

(1) $\{G_i\} \in \text{cov } \sigma(T)$;

(2) $X_i \subset \mathcal{D}_T$, if G_i is relatively compact ($G_i \in \mathcal{G}^K$);

(3) $X = \sum_i X_i$;

(4) $\sigma(T|X_i) \subset \overline{G}_i$ or, equivalently,

$\sigma(T|X_i) \subset G_i$ for all i.

Special properties of the invariant subspaces X_i induce special types of spectral decompositions. At center stage in this circle of ideas is the concept of *spectral decomposition property.* In the spirit of 1.1, we see that a spectral decomposition of X by T is formally operated by a map $E : G \to \text{Inv } T$, with the original family G providing the sets $\overline{G}_i$ to contain the spectra of the restrictions $T|E(G_i)$ and the final family Inv T supplying the summands X_i for the linear sum decomposition of X. While we defer the definition and the study of the map E until we reach Chapter IV, we indulge in a little discussion of our program.

In this work, we are primarily interested in extending the general spectral decomposition problem to the case of unbounded closed operators. When working with unbounded operators, the point at infinity of the one-point compactified complex plane $\mathbb{C}_\infty$ assumes a special role. If T is bounded then the resolvent operator is analytic at ∞ and

$$R(\infty;T) = \lim_{\lambda\to\infty} R(\lambda;T) = 0.$$

For unbounded T, $R(\cdot;T)$ has a singularity at ∞ and, for some purpose, the extended spectrum $\sigma_\infty(T) = \sigma(T) \cup \{\infty\}$ may be conveniently used. In many cases, however, when working within the topology of $\mathbb{C}$, we may still employ the ordinary spectrum $\sigma(T)$ of an unbounded T.

The basic requirement imposed by any spectral theory is the existence of proper invariant subspaces. A proper invariant subspace Y under T may be a summand of the spectral decomposition. Also Y produces the restriction $T|Y$ and the coinduced operator T/Y on the quotient space X/Y. Certain properties of $T|Y$ and T/Y may characterize the invariant subspace Y. The general and some special properties of invariant subspaces will be examined in terms of restriction and coinduced operators.

Some properties of bounded operators can be carried over to the unbounded case. Thus, in a few instances, the bounded case techniques can be adapted to unbounded operators. Most of the times, however, the proofs concerning unbounded operators are intrinsically different.

Originally, the concept of *decomposable operator* has been defined and its theory developed for bounded operators on a Banach space ([Fo.1963] and a wide variety of other papers). An extension of this concept to the unbounded case [V.1969; 1971,a] as well as to operators on a Fréchet space [V.1971] has been achieved by separating the part of the spectrum on which T failed to have the *single valued extension property.*

The minimal closed subset S of $\sigma_\infty(T)$ on whose complement T has the single valued extension property, in the sense that for any analytic function $f : \omega(\subset S^c) \to \mathcal{D}_T$, $(\lambda - T)f(\lambda) = 0$ on $\omega \cap \mathbb{C}$ implies $f(\lambda) = 0$ on ω, is called the *spectral residuum* of T. An open S-*cover* of $\sigma_\infty(T)$ has all but one member G_S disjoint from S. The pathology lurking in an S-cover is brought out in the pertinent spectral decomposition, by the invariant subspace that corresponds to G_S. The theory of these S- (or *residually*-) *decomposable operators* gives an interesting insight into the structure of some non-decomposable operators (e.g. [B.1975], [N.1979; 1979-1980; 1980], [T.1983], [V.1969; 1971; 1971,a; 1982], [W.1984], [W-Li.1984]).

To reach our targets, we shall follow a different path. The key ingredient in our approach is the spectral decomposition property [E-L.1978]. This property endows a closed operator with the single valued extension property [E.1980,a] and in the light of some subsequent works as [A.1979], [L.1981], [N.1978] and [Sh.1979], it gives a simpler and more natural interpretation to the concept of bounded decomposable operator. How much remains true if we drop the assumption of boundedness? The answer will arise from the spectral manifold discernibly at work with bounded and unbounded operators. What we loose is the property of the spectral manifold to be {0} at the empty set and this fact distinguishes between the operators with the spectral decomposition property which are decomposable and which are not.

The substance of the spectral theory suggests that a certain duality exists between an operator and its conjugate. It is a major topic of this work to explore the spectral duality of unbounded closed operators.

The use of certain spectral constructs, such as (*pre-*) *spectral capacity* and (*pre-*) *spectral resolvent* simplifies certain proofs and gives new characterizations to operators which possess some specific spectral properties. The interplay between various spectral resolvents and contingent properties of the operators will answer some questions and will open new problems.

For a deeper analysis, we shall frequently use the *contour integral* of a vector-valued function. An open $\Delta \subset \mathbb{C}$ is a *Cauchy domain* if it has a finite number of components and its boundary $\Gamma = \partial\Delta$ is a positively oriented finite system of closed, mutually nonintersecting rectifiable Jordan arcs. Γ will be referred to as an *admissible contour*. If a Cauchy domain Δ is a neighborhood of a set $S \subset \mathbb{C}$, we shall simply say that Δ is a *Cauchy neighborhood* of S. A set $H \subset \mathbb{C}$ is referred to as a *neighborhood of* ∞, in symbols $H \in V_\infty$, if the closure of its complement H^c is compact in $\mathbb{C}$. An open set ω is a neighborhood of $S \cup \{\infty\}$ if it is a neighborhood of both S and ∞. Without loss of generality, we assume that for $S \subset \mathbb{C}$, any $\{G_i\}_{i=0}^n \in \text{cov } S$ has, at most, one unbounded set G_0.

We denote by A_T the class of functions $f : \omega_f(\subset \mathbb{C}) \to \mathbb{C}$ which are locally analytic on a neighborhood ω_f of $\sigma_\infty(T)$ and regular at ∞. For $f \in A_T$, we write $f(\infty) = \lim_{\lambda\to\infty} f(\lambda)$. For the use of the contour integral it will be *implicitly assumed* that $\rho(T) \neq \emptyset$.

The functional calculus is established by the algebraic homomorphism $f \to f(T)$ between the algebra A_T and the *Banach algebra* B(X) of bounded linear operators which map X into X. If $f \in A_T$, then

$$f(T) = f(\infty) + \frac{1}{2\pi i}\int_{\partial\Delta} f(\lambda)R(\lambda;T)d\lambda,$$

where Δ is any Cauchy neighborhood of $\sigma_\infty(T)$ with $\bar{\Delta} \subset \omega_f$. If $f(\infty) = 0$, the range of f(T) is contained in $\mathcal{D}_T$.

It is an open question whether every bounded and unbounded closed operator on a Banach space has a proper invariant subspace. In fact, in terms of reducing subspaces, the unilateral shift operator on ℓ^2 is a counterexample [H.1951], [H.1967]. The existence of proper invariant subspaces for *subnormal operators* was proved by S. Brown [Bro.1979], (see also [St.1979]) and some extensions were obtained in [St.1980], [Ap.1980] and [Ap.1981]. In defining the spectral decomposition property of the given closed operator T, it will be assumed the existence of the necessary number of invariant subspaces with the required spectral property (as mentioned in the first paragraph of this section). The fact that we are not working in void will become soon evident by the wealth of the spectral maximal and T-bounded spectral maximal spaces, as constructive elements of the spectral decomposition problem.

§.2. THE SINGLE VALUED EXTENSION PROPERTY

The spectral decomposition problem cannot bé properly studied without the single valued extension property (SVEP). This property has a profound effect on both the operators and the invariant subspaces involved in a spectral decomposition.

2.1. DEFINITION. A linear operator $T : \mathcal{D}_T(\subset X) \to X$ is said to have the SVEP if, for every analytic[†] function $f : \omega_f \to \mathcal{D}_T$ defined on an open $\omega_f \subset \mathbb{C}$, the condition $(\lambda-T)f(\lambda) \equiv 0$ implies $f(\lambda) \equiv 0$.

Equivalently, for each $x \in X$, any two analytic functions $f : \omega_f \to \mathcal{D}_T$, $g : \omega_g \to \mathcal{D}_T$, satisfying condition

$$(\lambda-T)f(\lambda) = (\lambda-T)g(\lambda) = x \quad \text{on} \quad \omega_f \cap \omega_g,$$

agree on $\omega_f \cap \omega_g$. When this property holds, the union of the domains ω_f of all $\mathcal{D}_T$-valued analytic functions f, which identically verify equation

$$(2.1) \qquad (\lambda-T)f(\lambda) = x,$$

is called the *local resolvent set* and is denoted by $\rho(x,T)$. The SVEP implies the existence of a (maximally extended) analytic function $x_T(\cdot)$, or $x(\cdot)$ if T is understood, referred to as *local resolvent*, which maps $\rho(x,T)$ into $\mathcal{D}_T$ and identically verifies (2.1). The *local spectrum* $\sigma(x,T)$, defined as the complement in $\mathbb{C}$ of $\rho(x,T)$, is the set of singularities of $x(\cdot)$.

If not specified otherwise, we shall henceforth assume that the given T is a *closed operator*.

2.2. THEOREM. Given T, for every $x \in X$ and $\lambda_0 \in \mathbb{C}$, the following assertions are equivalent:

(I) there is a neighborhood δ of λ_0 and an analytic function $f : \delta \to \mathcal{D}_T$ verifying equation (2.1) on δ;

(II) there are numbers $M > 0$, $R > 0$ and a sequence $\{a_n\}_{n=0}^{\infty} \subset \mathcal{D}_T$ with the following properties:

$$(2.2) \quad \text{(a)} \ (\lambda_0-T)a_0 = x; \quad \text{(b)} \ (\lambda_0-T)a_{n+1} = a_n; \quad \text{(c)} \ \|a_n\| \le MR^n, \ n \in Z^+.$$

PROOF. (I) => (II): We may assume that

$$\delta = \{\lambda : |\lambda-\lambda_0| < r\} \quad \text{for some} \quad r > 0.$$

[†]the term "analytic" will be indistinguishably used for "locally analytic".

Let

(2.3) $$f(\lambda) = \sum_{n=0}^{\infty} a_n(\lambda_0-\lambda)^n, \quad \lambda \in \delta$$

be the power series expansion of f. By decreasing r, we may assume that (2.3) holds on $\overline{\delta}$. Then, for the radius r of $\partial\delta$, $\|a_n r^n\| \to 0$ as $n \to \infty$. Hence there is $M > 0$ such that

(2.4) $$\|a_n\| r^n \leq M, \quad n \in Z^+.$$

For $R = r^{-1}$, (2.4) implies (2.2,c). By making $\lambda = \lambda_0$ in (2.1) and (2.3), one obtains (2.2,a). Furthermore, it follows from (2.3) that

$$a_n = -\frac{1}{2\pi i}\int_{\partial\delta} \frac{f(\lambda)}{(\lambda_0-\lambda)^{n+1}}\, d\lambda, \quad n \in Z^+.$$

In view of (2.1), one can write

$$(\lambda_0-T)f(\lambda) = (\lambda_0-\lambda)f(\lambda) + (\lambda-T)f(\lambda) = (\lambda_0-\lambda)f(\lambda) + x.$$

Since T is closed, one obtains $a_n \in \mathcal{D}_T$ $(n \in Z^+)$ and

$$(\lambda_0-T)a_{n+1} = -\frac{1}{2\pi i}\int_{\partial\delta} \frac{(\lambda_0-T)f(\lambda)}{(\lambda_0-\lambda)^{n+2}}\, d\lambda$$

$$= -\frac{1}{2\pi i}\int_{\partial\delta} \frac{f(\lambda)}{(\lambda_0-\lambda)^{n+1}}\, d\lambda - \frac{1}{2\pi i}\int_{\partial\delta} \frac{x}{(\lambda_0-\lambda)^{n+2}}\, d\lambda = a_n.$$

This proves (2.2,b).

(II) => (I): In view of (2.2,c), the series (2.3) defines a function f, analytic on

$$\delta = \{\lambda : |\lambda-\lambda_0| < R^{-1}\}.$$

Thus, for

$$f_k(\lambda) = \sum_{n=0}^{k} a_n(\lambda_0-\lambda)^n, \quad \lambda \in \delta, \quad k \in \mathbb{N},$$

with the help of (2.2,a), one obtains

$$(\lambda-T)f_k(\lambda) = \sum_{n=0}^{k} (\lambda-T)a_n(\lambda_0-\lambda)^n = \sum_{n=0}^{k} (\lambda_0-T)a_n(\lambda_0-\lambda)^n$$

$$-\sum_{n=0}^{k} a_n(\lambda_0-\lambda)^{n+1} = x + \sum_{n=1}^{k} a_{n-1}(\lambda_0-\lambda)^n - \sum_{n=0}^{k} a_n(\lambda_0-\lambda)^{n+1}$$

$$= x - a_k(\lambda_0-\lambda)^{k+1}.$$

For every $\lambda \in \delta$, $f_k(\lambda) \to f(\lambda)$ and $a_k(\lambda_0-\lambda)^{k+1} \to 0$, as $k \to \infty$. Since T is closed, $f(\lambda) \in \mathcal{D}_T$ and $(\lambda-T)f(\lambda) = x$, for all $\lambda \in \delta$. □

2.3. COROLLARY. T does not have the SVEP iff there exists $\lambda_0 \in \mathbb{C}$ and there are numbers $M > 0$, $R > 0$ and a sequence $\{a_n\}_{n=0}^{\infty} \subset \mathcal{D}_T$ such that

(2.5) $(\lambda_0-T)a_0 = 0$; $(\lambda_0-T)a_{n+1} = a_n$; $\|a_n\| \leq MR^n$ $(n \in Z^+)$; $a_n \neq 0$ for some n.

PROOF. T does not have the SVEP iff, for some analytic function $f : \omega_f \to \mathcal{D}_T$ and $\lambda_0 \in \omega_f$, there is a neighborhood δ of λ_0 such that

(2.6) $$(\lambda-T)f(\lambda) \equiv 0 \quad \text{and} \quad f(\lambda) \not\equiv 0 \quad \text{on } \delta.$$

In view of 2.2, the situation described by (2.6) occurs iff conditions (2.5) hold. □

2.4. COROLLARY. T does not have the SVEP if there is $\lambda_0 \in \mathbb{C}$ such that λ_0-T is surjective but not injective.

PROOF. Assume that λ_0-T is surjective but not injective. Since λ_0-T is closed, it follows from the open mapping theorem that there is $R > 0$ such that, for each $y \in X$, there is $x \in \mathcal{D}_T$ satisfying conditions

$$(\lambda_0-T)x = y, \quad \|x\| \leq R\|y\|.$$

Since λ_0-T is not injective, one can choose $a_0 \in \mathcal{D}_T$ with $\|a_0\| = 1$ and $(\lambda_0-T)a_0 = 0$. For each $n \in \mathbb{N}$, let $a_n \in \mathcal{D}_T$ satisfy conditions

$$(\lambda_0-T)a_n = a_{n-1}, \quad \|a_n\| \leq R\|a_{n-1}\|.$$

Then (2.5) holds for $M = 1$ and hence T does not have the SVEP. □

2.5. COROLLARY. Let T have the SVEP. Then $\lambda \in \sigma(T)$ iff $\lambda-T$ is not surjective.

2.6. PROPOSITION. If T has the SVEP then the following properties hold:

(I) $\sigma(x+y,T) \subset \sigma(x,T) \cup \sigma(y,T)$; $x,y \in X$;

(II) $ax(\lambda) + by(\lambda) = (ax+by)(\lambda)$; $a,b \in \mathbb{C}$; $x,y \in X$; $\lambda \in \rho(x,T) \cap \rho(y,T)$;

(III) $\sigma(Ax,T) \subset \sigma(x,T)$ for every $A \in B(X)$ which commutes with T;

(IV) $\sigma(Tx,T) \subset \sigma(x,T)$ and $(Tx)(\lambda) = Tx(\lambda)$, $x \in \mathcal{D}_T$, $\lambda \in \rho(x,T)$;

(V) $\sigma[x(\lambda),T] = \sigma(x,T)$; $x \in X$, $\lambda \in \rho(x,T)$;

(VI) for any $A \in B(X)$ with the SVEP, $\sigma(x,A) = \emptyset$ iff $x = 0$.

PROOF. Properties (I), (II), (III) and (VI) can be proved as in the bounded case (e.g. [Du-S.1971; 1, XVI. 2.1, 2.2]).

(IV): It follows from

$$(\lambda-T)x(\lambda) = x, \quad x \in \mathcal{D}_T, \lambda \in \rho(x,T)$$

that $Tx(\lambda) \in \mathcal{D}_T$ and that $Tx(\cdot)$ is analytic on $\rho(x,T)$. Then

$$(\lambda-T)Tx(\lambda) = T(\lambda-T)x(\lambda) = Tx$$

implies (IV).

(V): Given $x \in X$, for every $\lambda \in \rho(x,T)$, there is an analytic function $\xi_\lambda : \rho[x(\lambda),T] \to \mathcal{D}_T$ verifying equation

$$(\mu-T)\xi_\lambda(\mu) = x(\lambda) \quad \text{on} \quad \rho[x(\lambda),T]. \tag{2.7}$$

Since, for $\lambda \in \rho(x,T)$, $x(\lambda) \in \mathcal{D}_T$, (2.7) implies that $T\xi_\lambda(\mu) \in \mathcal{D}_T$ and

$$(\mu-T)(\lambda-T)\xi_\lambda(\mu) = (\lambda-T)(\mu-T)\xi_\lambda(\mu) = (\lambda-T)x(\lambda) = x.$$

Since $(\lambda-T)\xi_\lambda(\mu) = (\lambda-\mu)\xi_\lambda(\mu) + x(\lambda)$ is analytic on $\rho[x(\lambda),T]$, we have $\mu \in \rho(x,T)$. Thus $\sigma(x,T) \subset \sigma[x(\lambda),T]$.

Conversely, for $\lambda \in \rho(x,T)$, define the analytic function $g_\lambda : \rho(x,T) \to X$, by

$$g_\lambda(\mu) = \begin{cases} -\dfrac{x(\mu) - x(\lambda)}{\mu-\lambda}, & \text{if } \mu \neq \lambda; \\ -x'(\lambda), & \text{if } \mu = \lambda. \end{cases}$$

For $\mu \neq \lambda$, we have $g_\lambda(\mu) \in \mathcal{D}_T$ and

$$(\mu-T)g_\lambda(\mu) = -\frac{x}{\mu-\lambda} + x(\lambda) + \frac{x}{\mu-\lambda} = x(\lambda). \tag{2.8}$$

T being closed, by letting $\mu \to \lambda$, one has $x'(\lambda) \in \mathcal{D}_T$ and hence (2.8) also holds for $\mu = \lambda$. Consequently, $\sigma[x(\lambda),T] \subset \sigma(x,T)$ and (V) follows. □

2.7. PROPOSITION. If T has the SVEP, then

$$\sigma(T) = \bigcup \{\sigma(x,T) : x \in X\}.$$

PROOF. Let $\lambda \in \mathbb{C} - \bigcup \{\sigma(x,T) : x \in X\}$. For every $x \in X$, $(\lambda-T)x(\lambda) = x$ implies that $\lambda-T$ is surjective. Then $\lambda \in \rho(T)$, by 2.5. Consequently, $\sigma(T) \subset \bigcup \{\sigma(x,T) : x \in X\}$. The opposite inclusion is obvious. □

If T has the SVEP, then 2.6 implies that, for every $H \subset \mathbb{C}$,

$$X(T,H) = \{x \in X : \sigma(x,T) \subset H\} \tag{2.9}$$

is a linear manifold in X. Moreover, if X(T,H) is closed then, by 2.6 (III) and (IV), X(T,H) is a *hyperinvariant* subspace under T (i.e. X(T,H) is invariant under every $A \in B(X)$ which commutes with T). The linear manifold (2.9) is called a *spectral manifold* (of T).

The SVEP is stable under functional calculus and the proof for an unbounded T follows by lines similar to that for a bounded operator [C-Fo.1967,1968].

2.8. PROPOSITION. Let T be such that $\rho(T) \neq \emptyset$. If T has the SVEP then, for each $f \in A_T$, f(T) has the same property. Conversely, if $f \in A_T$ is nonconstant on every component of its domain and if f(T) has the SVEP, then T has the SVEP.

The localized version of the spectral mapping theorem [Ap.1968] or [Bar-Ka.1973] has its extension to unbounded operators as given in [V.1982, IV. Theorem 3.12]. See also [Ho.1983,a].

2.9. PROPOSITION. Given T, let $f \in A_T$ be nonconstant on every component of its domain. Then, for every $x \in X$,

$$\sigma[x,f(T)] = f[\sigma_\infty(x,T)].$$

2.10. COROLLARY. Given T, let $f \in A_T$ be nonconstant on every component of its domain. If T has the SVEP then, for every set $H \subset \mathbb{C}$,

$$X[f(T),H] \subset X[T,f^{-1}(H)] \subset X[f(T),H \cup f(\{\infty\})]. \tag{2.10}$$

In particular, if $f(\infty) \in H$, then

$$X[f(T),H] = X[T,f^{-1}(H)]. \tag{2.11}$$

PROOF. By 2.9, for every $x \in X[f(T),H]$, we have

$$f[\sigma(x,T)] \subset \sigma[x,f(T)] \subset H$$

and hence $\sigma(x,T) \subset f^{-1}(H)$. Thus $x \in X[T,f^{-1}(H)]$ and the first inclusion of (2.10) follows. Next, let $x \in X[T,f^{-1}(H)]$. Then $\sigma(x,T) \subset f^{-1}(H)$ and (2.9) implies

$$\sigma[x,f(T)] = f[\sigma_\infty(x,T)] \subset f[f^{-1}(H) \cup \{\infty\}] \subset H \cup f(\{\infty\}).$$

Hence, the second inclusion of (2.10) is obtained. Now (2.11) is a direct consequence of $f(\infty) \in H$ and (2.10). □

§.3. INVARIANT SUBSPACES. GENERAL PROPERTIES.

To develop the constructive elements of the spectral decomposition of X, we devote this section to a spectral-theoretic study of invariant subspaces. A subspace Y of X is *invariant* under T, in symbols, $Y \in \text{Inv } T$, if $T(Y \cap \mathcal{D}_T) \subset Y$. An invariant subspace Y produces two operators: the restriction $T|Y$ and the coinduced $\hat{T} = T/Y$ by T on the quotient space X/Y. The latter has the domain

$$\mathcal{D}_{\hat{T}} = \{\hat{x} \in X/Y : \hat{x} \cap \mathcal{D}_T \neq \emptyset\}$$

and, for $\hat{x} \in \mathcal{D}_{\hat{T}}$, $x \in \hat{x} \cap \mathcal{D}_T$, we define $\hat{T}\hat{x} = (Tx)^\wedge$.

3.1. PROPOSITION. Given T and $Y \in \text{Inv } T$, consider the following conditions:

(3.1) $$\sigma(T) \cup \sigma(T|Y) \neq \mathbb{C};$$

(3.2) $$\hat{T} = T/Y \text{ is closed on } X/Y.$$

Then (3.1) implies (3.2) and either of them produces the following inclusions

(3.3) $$\sigma(\hat{T}) \subset \sigma(T) \cup \sigma(T|Y);$$

$$\sigma(T|Y) \subset \sigma(T) \cup \sigma(\hat{T});$$

$$\sigma(T) \subset \sigma(\hat{T}) \cup \sigma(T|Y).$$

PROOF. Assume (3.1) and let $\lambda \in \rho(T) \cap \rho(T|Y)$ be arbitrary. For any $x \in Y$, we have $R(\lambda;T)x = R(\lambda;T|Y)x \in Y$ and hence Y is invariant under $R_\lambda = R(\lambda;T)$. Let $\hat{R}_\lambda$ be the coinduced operator by R_λ on X/Y. The identities

$$(\lambda-T)R_\lambda x = x,\ x \in X;\quad R_\lambda(\lambda-T)x = x,\ x \in \mathcal{D}_T$$

give rise to

(3.4) $$(\lambda-\hat{T})\hat{R}_\lambda\hat{x} = \hat{x},\ \hat{x} \in X/Y;\quad \hat{R}_\lambda(\lambda-\hat{T})\hat{x} = \hat{x},\ \hat{x} \in \mathcal{D}_{\hat{T}}.$$

It follows from (3.4) that $\hat{R}_\lambda$ is the inverse of $\lambda-\hat{T}$. Since $\hat{R}_\lambda$ is bounded and defined on X/Y, it is closed and hence $\hat{T}$ is closed. Furthermore, by

(3.4), $\lambda \in \rho(\hat{T})$ and this implies (3.3). The remainder of the proof is routine and we omit it. □

3.2. PROPOSITION. Given T, let $X_0, X_1, Y \in \mathrm{Inv}\, T$ satisfy the following conditions

$$X = X_0 + X_1, \quad X_1 \subset \mathcal{D}_T \cap Y; \tag{3.5}$$

$$\sigma(T|X_0) \subset F, \quad \sigma(T|X_0 \cap Y) \subset F, \tag{3.6}$$

for some closed $F \subsetneqq \mathbb{C}$. Then $\hat{T} = T/Y$ is closed. Moreover, if $\tilde{T} = (T|X_0)/Y \cap X_0$ (i.e. $\tilde{T}$ is the coinduced operator by $T|X_0$ on $X_0/Y \cap X_0$), then $\hat{T}$ and $\tilde{T}$ are similar and hence

$$\sigma(\hat{T}) = \sigma(\tilde{T}). \tag{3.7}$$

PROOF. Since, by (3.6),

$$\sigma(T|X_0) \cup \sigma(T|Y \cap X_0) \subset F \neq \mathbb{C},$$

$\tilde{T}$ is closed, by 3.1. Next, we show that $\hat{T}$ and $\tilde{T}$ are similar. In view of (3.5), each $x \in \mathcal{D}_T$ has a representation

$$x = x_0 + x_1 \quad \text{with} \quad x_i \in X_i, \quad i=0,1.$$

Since $x_1 \in \mathcal{D}_T$, we have $x_0 \in \mathcal{D}_T$. Therefore, $x_1 \in Y \cap \mathcal{D}_T$ and $x_0 \in X_0 \cap \mathcal{D}_T$. For $x \in X$, let $\hat{x} = x + Y \in X/Y$ and, for $x_0 \in X_0$, let $\tilde{x}_0 = x_0 + Y \cap X_0 \in X_0/Y \cap X_0$. The spaces X/Y, $X_0/Y \cap X_0$ are topologically isomorphic. Let $A : X/Y \to X_0/Y \cap X_0$, with $A\hat{x} = \tilde{x}_0$, be the topological isomorphism. For every $\hat{x} \in \mathcal{D}_{\hat{T}}$, there is $x \in \hat{x} \cap \mathcal{D}_T$ with $A\hat{x} = \tilde{x}_0 \in \mathcal{D}_{\tilde{T}}$. Conversely, for every $\tilde{x}_0 \in \mathcal{D}_{\tilde{T}}$, there is $x_0 \in \tilde{x}_0 \cap (X_0 \cap \mathcal{D}_T)$ and hence $\hat{x} \in \mathcal{D}_{\hat{T}}$. Consequently, $A\mathcal{D}_{\hat{T}} = \mathcal{D}_{\tilde{T}}$. For each $\hat{x} \in \mathcal{D}_{\hat{T}}$, one obtains

$$A\hat{T}\hat{x} = A(Tx)^{\wedge} = (Tx_0)^{\sim} = \tilde{T}\tilde{x}_0 = \tilde{T}A\hat{x}.$$

Hence $\hat{T}$ and $\tilde{T}$ are similar. Consequently, $\hat{T}$ is closed and (3.7) holds. □

3.3. COROLLARY. Given T, let $X_0, X_1 \in \mathrm{Inv}\, T$ be such that

$$X = X_0 + X_1, \quad X_1 \subset \mathcal{D}_T; \quad \sigma(T|X_0) \subset F, \quad \sigma(T|X_0 \cap X_1) \subset F,$$

for some closed $F \subsetneqq \mathbb{C}$. Then T/X_1 is closed and

$$\sigma(T/X_1) \subset \sigma(T|X_0) \cup \sigma(T|X_0 \cap X_1).$$

PROOF. For $Y = X_1$, the corollary is a direct consequence of 3.2. □

3.4. PROPOSITION. Given T, let $X_0, X_1 \in \operatorname{Inv} T$ be such that

$$X = X_0 + X_1, \quad X_1 \subset \mathcal{D}_T.$$

Then $\hat{T} = T/X_0$ is bounded and

$$\sigma(\hat{T}) \subset \sigma(T|X_1) \cup \sigma(T|X_0 \cap X_1). \tag{3.8}$$

PROOF. Let $x \in X$ and put $\hat{x} = x + X_0$. It follows from

$$x = x_0 + x_1, \quad x_0 \in X_0, \quad x_1 \in X_1 \subset \mathcal{D}_T,$$

that $x_1 \in \hat{x} \cap \mathcal{D}_T$. Therefore, $\hat{x} \cap \mathcal{D}_T \neq \emptyset$. Thus $\hat{x} \in \mathcal{D}_{\hat{T}}$ and hence $\mathcal{D}_{\hat{T}} = X/X_0$. Since the quotient spaces X/X_0 and $X_1/X_0 \cap X_1$ are topologically isomorphic, $\hat{T}$ and $\tilde{T} = (T|X_1)/X_0 \cap X_1$ are similar, by 3.2. Since, by 3.2, $\tilde{T}$ is closed, so is $\hat{T}$. Thus $\hat{T}$ is bounded, by the closed graph theorem. Since $\sigma(\hat{T}) = \sigma(\tilde{T})$, (3.8) follows from (3.3). □

We recall that if $T : \mathcal{D}_T(\subset X) \to Y$ and $A : \mathcal{D}_A(\subset Y^*) \to X^*$ are adjoint to each other and one of them is densely defined, then the other is closable (e.g. [Kat.1966, III.5. Theorem 5.28]). The following proposition gives a condition for a coinduced T/Y to be closable on X/Y.

3.5. PROPOSITION. Let T be densely defined and let $Y \in \operatorname{Inv} T$ be such that $\overline{Y \cap \mathcal{D}_T} = Y$. If $T^*|Y^a$ is densely defined then T/Y is closable. Moreover,

$$(T/Y)^* = T^*|Y^a, \quad (Y^a \text{ is the annihilator of } Y \text{ in } X^*).$$

PROOF. The fact that $Y^a \in \operatorname{Inv} T^*$, follows easily. Indeed, for $x \in Y \cap \mathcal{D}_T$ and $x^* \in Y^a \cap \mathcal{D}_{T^*}$, we have $0 = \langle Tx, x^* \rangle = \langle x, T^*x^* \rangle$. Since $\overline{Y \cap \mathcal{D}_T} = Y$, one obtains $T^*x^* \in Y^a$.

Y^a can be viewed as the dual of X/Y, under the isometric isomorphism $(X/Y)^* \to Y^a$. For convenience, we make no distinction between Y^a and $(X/Y)^*$, and denote by $\langle \hat{x}, x^* \rangle$ the linear functional $x^* \in Y^a$ on X/Y. For $\hat{x} \in \mathcal{D}_{T/Y}$, $x \in \hat{x} \cap \mathcal{D}_T$, $y \in Y$ and $x^* \in Y^a \cap \mathcal{D}_{T^*}$, one obtains

$$\begin{aligned} \langle (T/Y)\hat{x}, x^* \rangle &= \langle (Tx)^\wedge, x^* \rangle = \langle Tx + y, x^* \rangle = \langle Tx, x^* \rangle \\ &= \langle x, T^*x^* \rangle = \langle x + y, T^*x^* \rangle = \langle \hat{x}, T^*x^* \rangle. \end{aligned} \tag{3.9}$$

Consequently, T/Y and $T^*|Y^a$ are conjugate to each other and since $T^*|Y^a$ is densely defined, T/Y is closable.

To prove the second statement, note that $\overline{\mathcal{D}}_T = X$ implies that $\overline{\mathcal{D}}_{T/Y} = X/Y$. Thus, T/Y is densely defined and hence the conjugate $(T/Y)^*$ exists. If $G(\cdot)$ denotes the graph of an operator and $VG(\cdot)$ is the inverse graph (i.e. the mapping $V : X \times X \to X \times X$ is defined by $V(x,y) = (-y,x)$), then it follows from (3.9) that

$$VG(T^*|Y^a) \subset G(T/Y)^a = VG[(T/Y)^*]$$

and hence $(T/Y)^* \supset T^*|Y^a$. Now, let $x^* \in \mathcal{D}_{(T/Y)^*}$. For $x \in \mathcal{D}_T$ and $y \in Y$,

$$(3.10) \qquad \langle (T/Y)\hat{x}, x^*\rangle = \langle \hat{x}, (T/Y)^*x^*\rangle = \langle x+y, (T/Y)^*x^*\rangle = \langle x, (T/Y)^*x^*\rangle.$$

Thus, for every $x^* \in \mathcal{D}_{(T/Y)^*}$, $\langle (T/Y)\hat{x}, x^*\rangle$ is a bounded linear functional on $\mathcal{D}_T$ and hence $x^* \in \mathcal{D}_{T^*}$. Furthermore, $x^* \in \mathcal{D}_{(T/Y)^*} \subset Y^a$ and hence $x^* \in Y^a \cap \mathcal{D}_{T^*}$. We have

$$(3.11) \qquad \langle (T/Y)\hat{x}, x^*\rangle = \langle Tx + y, x^*\rangle = \langle Tx, x^*\rangle = \langle x, T^*x^*\rangle.$$

It follows from (3.10) and (3.11) that $(T/Y)^* \subset T^*|Y^a$. □

3.6. PROPOSITION. Given T, let $Y \in \text{Inv } T$. Then
(I) for any component G of $\rho(T)$, either $G \subset \sigma(T|Y)$ or $G \subset \rho(T|Y)$;
(II) if $Y \subset \mathcal{D}_T$, then each unbounded component of $\rho(T)$ is contained in $\rho(T|Y)$.

PROOF. (I). Suppose that we simultaneously have

$$\sigma(T|Y) \cap G \neq \emptyset \quad \text{and} \quad \rho(T|Y) \cap G \neq \emptyset.$$

Then $\partial\sigma(T|Y) \cap G \neq \emptyset$ and hence there is $\lambda \in \mathbb{C}$ such that

$$\lambda \in \partial\sigma(T|Y) \cap G \subset \sigma_a(T|Y) \subset \sigma_a(T) \subset \sigma(T).$$

This, however, is a contradiction.

(II). Since $Y \subset \mathcal{D}_T$, $\sigma(T|Y)$ is compact. If G is an unbounded component of $\rho(T)$ such that $\sigma(T|Y) \cap G \neq \emptyset$, then $\partial\sigma(T|Y) \cap G \neq \emptyset$. Now, an argument similar to that used in the proof of (I), leads one to a contradiction. □

We proceed with some elementary but useful properties of a closed operator acting on a direct sum decomposition of the underlying Banach space.

3.7. LEMMA. Let X be the direct sum of two subspaces X_1, X_2,

$$X = X_1 \oplus X_2$$

and let $T_i : \mathcal{D}_{T_i}(\subset X_i) \to X_i$ $(i=1,2)$ be closed operators. The linear operator $T : \mathcal{D}_T(\subset X) \to X$, defined by

$$\mathcal{D}_T = \{x \in X : x = x_1 + x_2,\ x_i \in \mathcal{D}_{T_i},\ i=1,2\},$$

$$Tx = T_1x_1 + T_2x_2, \quad x \in \mathcal{D}_T,$$

is closed and

$$\sigma(T) = \sigma(T_1) \cup \sigma(T_2).$$

PROOF. The proof is routine and is omitted. □

The *reducibility* of an operator in terms of *spectral sets* is now appropriate to be recorded. The theory is known (e.g. [Ta-La.1980,V.9.]).

3.8. THEOREM. Given T, if for closed disjoint sets σ_0, σ_1 with σ_0 compact, one has

$$\sigma(T) = \sigma_0 \cup \sigma_1$$

then there exist $X_0, X_1 \in \text{Inv } T$ satisfying the following conditions:

(I) $X = X_0 \oplus X_1$;

(II) $X_1 \subset \mathcal{D}_T$, $\sigma(T|X_i) = \sigma_i$ (i=0,1).

3.9. COROLLARY. Given T, let $Y \in \text{Inv } T$ be such that $\sigma(T|Y)$ is compact. Then, there exist $\mathrm{T}, W \in \text{Inv } T$ with the following properties:

(i) $Y = \mathrm{T} \oplus W$;

(ii) $\mathrm{T} \subset \mathcal{D}_T$, $\sigma(T|\mathrm{T}) = \sigma(T|Y)$, $\sigma(T|W) = \emptyset$.

PROOF. For $\sigma_1 = \sigma(T|\mathrm{T})$, $\sigma_0 = \emptyset$, 3.8 applied to $T|Y$ gives rise to (i) and (ii). If Δ is a bounded Cauchy neighborhood of $\sigma(T|Y)$, the projection

$$Q = \frac{1}{2\pi i} \int_{\partial\Delta} R(\lambda; T|Y)d\lambda \tag{3.12}$$

produces $\mathrm{T} = QY$ and $W = (I_Y - Q)Y$, where I_Y is the identity in Y. □

3.10. LEMMA. Let X_1 and X_2 be subspaces of X such that

$$X = X_1 + X_2. \tag{3.13}$$

There is a constant $M > 0$ such that, for every $x \in X$ there is a representation

$$x = x_1 + x_2, \quad x_i \in X_i \ (i=1,2) \tag{3.14}$$

satisfying condition

$$\|x_1\| + \|x_2\| \le M\|x\|. \tag{3.15}$$

PROOF. Define the continuous map $P : X_1 \oplus X_2 \to X$, by

$$P(x_1 \oplus x_2) = x_1 + x_2,$$

equipped with the norm

$$\|x_1 \oplus x_2\| = \|x_1\| + \|x_2\|.$$

P is surjective, by (3.13). By the open mapping theorem, there is a constant $M > 0$ such that, for every $x \in X$ with (3.14), there exists

$$y = x_1 \oplus x_2 \in X_1 \oplus X_2$$

satisfying conditions

$$Py = x \quad \text{and} \quad \|y\| \leq M\,\|x\|.$$

Since $\|y\| = \|x_1\| + \|x_2\|$, (3.15) is obtained. □

3.11. PROPOSITION. Let X_1, X_2 be subspaces of X satisfying (3.13). If $f : \omega_f \to X$ is analytic on an open $\omega_f \subset \mathbb{C}$ then, for every $\lambda_0 \in \omega_f$, there is a neighborhood $\omega_0 (\subset \omega_f)$ of λ_0 and there are analytic functions $f_i : \omega_0 \to X_i$ $(i=1,2)$ such that

(3.16) $$f(\lambda) = f_1(\lambda) + f_2(\lambda) \quad \text{on } \omega_0.$$

PROOF. Put

$$\omega_0(\subset \omega_f) = \{\lambda : |\lambda - \lambda_0| < r\} \text{ for some } r > 0$$

and let

(3.17) $$f(\lambda) = \sum_{n=0}^{\infty} a_n(\lambda-\lambda_0)^n, \quad \{a_n\} \subset X,$$

be the power series expansion of f in ω_0. By decreasing r, we may assume that

(3.18) $$\sup_n r^n \|a_n\| < \infty.$$

By 3.10, for every $n \in Z^+$, there is a representation

(3.19) $$a_n = a_{1n} + a_{2n}, \quad a_{in} \in X_i \quad (i=1,2)$$

with

(3.20) $$\|a_{1n}\| + \|a_{2n}\| \leq M\|a_n\| \text{ for some constant } M > 0.$$

In view of (3.18) and (3.20), the series

(3.21) $$f_i(\lambda) = \sum_{n=0}^{\infty} a_{in}(\lambda-\lambda_0)^n, \quad i=1,2$$

converge and hence the functions f_i are analytic on ω_0. Consequently, (3.17), (3.19) and (3.21) give rise to (3.16). □

§.4. INVARIANT SUBSPACES. SPECIAL PROPERTIES.

The time has come to have a closer look at the invariant subspaces which occur most frequently in spectral decompositions.

4.1. DEFINITION. Given T, $Y \in \text{Inv } T$ is said to be a *ν-space* of T if

$$\sigma(T|Y) \subset \sigma(T).$$

A useful criterion for an invariant subspace to be a ν-space, with the proof [Sc.1959, Theorem 4] invariably valid in the unbounded case, is expressed by

4.2. PROPOSITION. Given T with $\rho(T) \neq \emptyset$, $Y \in \text{Inv } T$ is a ν-space of T iff $R(\lambda;T)Y \subset Y$ for all $\lambda \in \rho(T)$.

4.3. COROLLARY. Every hyperinvariant subspace under T is a ν-space of T.

PROOF. If $\rho(T) = \emptyset$, then every invariant subspace is a ν-space of T. Assume that $\rho(T) \neq \emptyset$ and let Y be a hyperinvariant subspace. Since, for each $\lambda \in \rho(T)$, $R(\lambda;T)$ commutes with T, the hypothesis on Y implies that $R(\lambda;T)Y \subset Y$. Thus, Y is a ν-space of T, by 4.2. □

4.4. COROLLARY. Given T, let Y be hyperinvariant under T and suppose that $\sigma(T|Y)$ is compact. Then, the subspaces $\mathcal{T}$ and W, as defined in 3.9, are hyperinvariant under T.

PROOF. Let $A \in B(X)$ commute with T. Then $Y \in \text{Inv } A$ and $A|Y$ commutes with $T|Y$. Moreover, for each $\lambda \in \rho(T|Y)$, $A|Y$ commutes with $R(\lambda;T|Y)$ and hence $A|Y$ commutes with the projection Q (3.12). Then $\mathcal{T} = QY$ and $W = (I_Y - Q)Y$ are invariant under $A|Y$. Thus, $\mathcal{T}$ and W are invariant under A. □

4.5. PROPOSITION. Let T have the SVEP and suppose that $X = \sum_{i=1}^{n} Y_i$. If each Y_i is a ν-space of T, then $\sigma(T) = \bigcup_{i=1}^{n} \sigma(T|Y_i)$.

PROOF. The hypothesis on the Y_i's implies that $\sigma(T) \supset \bigcup_{i=1}^{n} \sigma(T|Y_i)$. The opposite inclusion follows from 2.6 (I) and 2.7. □

If the Y_i's are the summands of a (weak) spectral decomposition of X by T, then T has a *decomposable spectrum* ([J.1977], [Ho.1982]).

4.6. PROPOSITION. If T has the SVEP and $Y \in \text{Inv } T$, then

$$\sigma(y,T) \subset \sigma(y,T|Y) \quad \text{for all} \quad y \in Y.$$

PROOF. Let $y \in Y$. For every $\lambda \in \rho(y,T|Y)$,

$$(\lambda-T)y_T(\lambda) = (\lambda-T|Y)y_{T|Y}(\lambda) = y$$

and hence $\rho(y,T|Y) \subset \rho(y,T)$. □

4.7. DEFINITION. Let T have the SVEP. Then $Y \in \text{Inv } T$ is said to be a *μ-space* of T if

$$\sigma(y,T) = \sigma(y,T|Y) \quad \text{for all} \quad y \in Y.$$

In view of 4.6, $Y \in \text{Inv } T$ is a μ-space of T iff

(4.1) $$\sigma(y,T) \supset \sigma(y,T|Y) \quad \text{for all} \quad y \in Y.$$

4.8. PROPOSITION. Let T have the SVEP. Then

(i) each μ-space of T is also a ν-space of T;

(ii) $Y \in \text{Inv } T$ is a μ-space of T iff

$$\{y(\lambda) : \lambda \in \rho(y,T),\ y \in Y\} \subset Y.$$

PROOF. (i): Let Y be a μ-space of T. With the help of 2.7, one obtains

$$\sigma(T|Y) = \cup\{\sigma(y,T|Y) : y \in Y\} = \cup\{\sigma(y,T) : y \in Y\}$$
$$\subset \cup\{\sigma(x,T) : x \in X\} = \sigma(T).$$

(ii): First, suppose that Y is a μ-space of T. Then, for all $y \in Y$ and $\lambda \in \rho(y,T) = \rho(y,T|Y)$, one has

$$y(\lambda) = y_T(\lambda) = y_{T|Y}(\lambda) \in Y.$$

Conversely, if for all $y \in Y$ and $\lambda \in \rho(y,T)$ we have $y(\lambda) \in Y$, then

$$(\lambda-T|Y)y(\lambda) = (\lambda-T)y(\lambda) = y$$

and hence $\rho(y,T) \subset \rho(y,T|Y)$. Thus Y is a μ-space of T, by (4.1). □

4.9. PROPOSITION. Let T have the SVEP. Then $Y \in \text{Inv } T$ is a μ-space of T iff, for every closed F,

(4.2) $$Y \cap X(T,F) = Y(T|Y,F).$$

For F closed, $Y(T|Y,F) \subset Y \cap X(T,F)$, by 4.6. If Y is a μ-space of T then, for $y \in Y \cap X(T,F)$, one has $\sigma(y,T|Y) = \sigma(y,T) \subset F$ and hence

$$Y \cap X(T,F) \subset Y(T|Y,F).$$

Conversely, assume that (4.2) holds and let $y \in Y$. Denote $F = \sigma(y,T)$ and obtain $y \in Y \cap X(T,F) = Y(T|Y,F)$. Therefore, $\sigma(y,T|Y) \subset F = \sigma(y,T)$ and hence Y is a μ-space of T, by (4.1). □

4.10. DEFINITION. Given T, $Y \in \operatorname{Inv} T$ is called an *analytically invariant* subspace under T if, for any analytic function $f : \omega_f \to \mathcal{D}_T$, the condition $(\lambda-T)f(\lambda) \in Y$ implies that $f(\lambda) \in Y$ on an open $\omega_f \subset \mathbb{C}$.

We write $AI(T)$ for the family of analytically invariant subspaces under T.

4.11. PROPOSITION. Every analytically invariant subspace Y under T is a ν-space of T. If, in addition T has the SVEP, then Y is a μ-space of T.

PROOF. Let $Y \in AI(T)$ and $y \in Y$. Since $y = (\lambda-T)R(\lambda;T)y \in Y$ on $\rho(T)$, $R(\lambda;T)y \in Y$ on $\rho(T)$ and hence Y is a ν-space of T, by 4.2. Moreover, if T has the SVEP then, for $y \in Y$ and $\lambda \in \rho(y,T)$, $(\lambda-T)y(\lambda) = y$ implies that $y(\lambda) \in Y$. Thus Y is a μ-space of T, by 4.8 (ii). □

4.12. PROPOSITION. Given T, let $Y \in AI(T)$ be such that $\sigma(T|Y)$ is compact. Then $\mathcal{T} \in AI(T)$ and, if T has the SVEP then $W \in AI(T)$, where $\mathcal{T}$, W were defined by 3.9.

PROOF. Let $f : \omega_f \to \mathcal{D}_T$ be analytic and satisfy condition

$$(\lambda-T)f(\lambda) \in \mathcal{T} \text{ on an open } \omega_f \subset \mathbb{C}. \tag{4.3}$$

Since $\mathcal{T} \subset Y$ and $Y \in AI(T)$, (4.3) implies that $f(\lambda) \in Y$ on ω_f. In view of 3.11, there are analytic functions $f_1 : \omega \to \mathcal{T}$, $f_2 : \omega \to W$ such that

$$f(\lambda) = f_1(\lambda) + f_2(\lambda) \text{ on an open } \omega \subset \omega_f.$$

Since $f(\omega) \subset \mathcal{D}_T$ and $f_1(\omega) \subset \mathcal{T} \subset \mathcal{D}_T$, it follows that $f_2(\omega) \subset \mathcal{D}_T$. Then (4.3) implies

$$(\lambda-T)f_1(\lambda) \in \mathcal{T}, \quad \lambda \in \omega;$$

$$(\lambda-T)f_2(\lambda) = 0, \quad \lambda \in \omega. \tag{4.4}$$

Since $\sigma(T|W) = \emptyset$, it follows from (4.4) that $f_2(\lambda) \equiv 0$ and hence $f(\lambda) = f_1(\lambda)$ on ω. Thus $f(\lambda) \in \mathcal{T}$ on ω_f, by analytic continuation. The proof of the second assertion of the proposition is left to the reader. □

4.13. PROPOSITION. Given $T \in B(X)$, suppose that $\sigma(T)$ is nowhere dense and does not separate the plane. Then, every $Y \in \operatorname{Inv} T$ is analytically invariant under T.

PROOF. Let $f : \omega_f \to X$ be analytic and satisfy condition

$$h(\lambda) = (\lambda-T)f(\lambda) \in Y \quad \text{on an open } \omega_f \subset \mathbb{C}.$$

Without loss of generality, we may assume that ω_f is connected. Since $\sigma(T)$ is nowhere dense, $\omega_f \cap \rho(T) \neq \emptyset$. Since $\sigma(T)$ does not separate the plane, Y is a ν-space of T. Then, for $\lambda \in \omega_f \cap \rho(T)$, 4.2 implies that $f(\lambda) = R(\lambda;T)h(\lambda) \in Y$ and hence $f(\lambda) \in Y$ on ω_f, by analytic continuation. □

Next, we extend a useful property [Fr.1973, Theorem 1] of analytically invariant subspaces to unbounded closed operators.

4.14. PROPOSITION. Given T, let $Y \in \text{Inv } T$ be such that $Y \subset \mathcal{D}_T$. Y is analytically invariant under T iff $\hat{T} = T/Y$ has the SVEP.

PROOF. First, assume that $\hat{T}$ has the SVEP and let $f : \omega_f \to \mathcal{D}_T$ be analytic and satisfy condition

$$(\lambda-T)f(\lambda) \in Y \quad \text{on an open } \omega_f \subset \mathbb{C}.$$

By the natural homomorphism $X \to X/Y$, we have

$$(\lambda-\hat{T})\hat{f}(\lambda) = \hat{0} \quad \text{on } \omega_f.$$

By the SVEP, $\hat{f}(\lambda) \equiv \hat{0}$ and hence $f(\lambda) \in Y$ for all $\lambda \in \omega_f$.

Conversely, assume that $Y \in AI(T)$. Let $\hat{f} : \omega_{\hat{f}} \to \mathcal{D}_{\hat{T}}$ be analytic and satisfy condition

$$(\lambda-\hat{T})\hat{f}(\lambda) = \hat{0} \quad \text{on an open } \omega_{\hat{f}} \subset \mathbb{C}. \tag{4.5}$$

Without loss of generality, we assume that $\omega_{\hat{f}}$ is connected. Let

$$\hat{f}(\lambda) = \sum_{n=0}^{\infty} \hat{a}_n(\lambda-\lambda_0)^n$$

be the power series expansion of $\hat{f}$ in a neighborhood δ of $\lambda_0 \in \omega_{\hat{f}}$. For each n, one can choose $a_n \in \hat{a}_n$ such that $\|a_n\| \leq \|\hat{a}_n\| + 1$. Then

$$\limsup_n \|a_n\|^{1/n} \leq \limsup_n \|\hat{a}_n\|^{1/n} + 1$$

and hence

$$f(\lambda) = \sum_{n=0}^{\infty} a_n(\lambda-\lambda_0)^n \in \hat{f}(\lambda)$$

is analytic on a neighborhood $\delta' \subset \delta \subset \omega_{\hat{f}}$ of λ_0. Since $\hat{f}(\lambda) \in \mathcal{D}_{\hat{T}}$, there is $h(\lambda) \in \hat{f}(\lambda) \cap \mathcal{D}_T$. Then $\hat{f}(\lambda) = h(\lambda) + Y \subset \mathcal{D}_T$ and hence

$f(\lambda) \in \hat{f}(\lambda) \subset \mathcal{D}_T$, for all $\lambda \in \delta'$. Then (4.5) implies

$$(\lambda - T)f(\lambda) \in Y \quad \text{on } \delta'$$

and by the hypothesis on Y, we have $f(\lambda) \in Y$ on δ'. Thus, $\hat{f}(\lambda) = \hat{0}$ on δ' and hence $\hat{f}(\lambda) = \hat{0}$ on $\omega_{\hat{f}}$, by analytic continuation. □

The following lemma which appeared in [N.1981, Lemma 3.2] has many useful applications.

4.15. LEMMA. Given T, let $Y \in \text{Inv}\, T$ with $Y \subset \mathcal{D}_T$ be such that $\hat{T} = T/Y$ is closed in X/Y. Suppose that, for $\hat{x} \in X/Y$ and $z \in \mathbb{C}_\infty$, there is a neighborhood V of z and an analytic function $\hat{g} : V \to \mathcal{D}_{\hat{T}}$ satisfying the following condition

$$(\lambda - \hat{T})\hat{g}(\lambda) = \hat{x} \quad \text{for } \lambda \in V \cap \mathbb{C}.$$

Then, there is another neighborhood $V' \subset V$ of z and an analytic function $h : V' \to \mathcal{D}_T$ such that $\hat{h}(\lambda) = \hat{g}(\lambda)$ on V' and $(\lambda - T)h(\lambda)$ is analytic on V'.

PROOF. Let D denote the linear manifold $\mathcal{D}_T$ endowed with the graph norm

$$\|x\|_T = \|x\| + \|Tx\|.$$

T being closed, D is a Banach space and so is D/Y with respect to the usual norm $\|\cdot\|_{D/Y}$ of the quotient space. $D/Y = \mathcal{D}_{\hat{T}}$ can also be endowed with the graph norm $\|\hat{x}\|_{\hat{T}} = \|\hat{x}\| + \|\hat{T}\hat{x}\|$ and since $\hat{T}$ is closed, D/Y is a Banach space with respect to the graph norm $\|\cdot\|_{\hat{T}}$. For any $\hat{x} \in \mathcal{D}_{\hat{T}}$ and all $x \in \hat{x}$, we have

$$\|\hat{x}\|_{\hat{T}} = \|\hat{x}\| + \|\hat{T}\hat{x}\| = \inf_{y \in Y} \|x+y\| + \inf_{w \in Y} \|Tx+w\| \le \inf_{y \in Y} \|x+y\|$$

$$+ \inf_{y \in Y} \|Tx+Ty\| \le \inf_{y \in Y} \{ \|x+y\| + \|T(x+y)\| \} = \|\hat{x}\|_{D/Y}.$$

Since D/Y is complete under either norm $\|\cdot\|_{\hat{T}}$, $\|\cdot\|_{D/Y}$, it follows from the open mapping theorem that the two norms are equivalent.

For $\lambda \in V \cap \mathbb{C}$, we have

$$\hat{T}\hat{g}(\lambda) = \lambda\hat{g}(\lambda) - \hat{x}. \tag{4.6}$$

We examine the two possible cases: (a) z is finite and (b) $z = \infty$. In case (a), we may assume that $V \subset \mathbb{C}$. Then $\hat{T}\hat{g}(\cdot)$ is analytic and hence $\hat{g}$ is analytic on V under the norm $\|\cdot\|_{\hat{T}}$ or, equivalently, under the norm $\|\cdot\|_{D/Y}$. By [V.1971, Lemma 2.1], there is a neighborhood $V' \subset V$ of z and an analytic function $h : V' \to D$ such that $\hat{h}(\lambda) = \hat{g}(\lambda)$ on V'. Since $V' \subset \mathbb{C}$ and h is analytic under the norm $\|\cdot\|_T$, $(\lambda-T)h(\lambda)$ is analytic on V'. In case (b), (4.6) rewritten as

$$\hat{T}\,\frac{\hat{g}(\lambda)}{\lambda} = \hat{g}(\lambda) - \frac{\hat{x}}{\lambda},$$

implies that $\hat{g}(\infty) = \hat{0}$. Thus $\lambda\hat{g}(\lambda)$ is analytic and hence so is $\hat{T}\hat{g}(\lambda)$ on V. Consequently, $\hat{g}$ is analytic on V under the norm $\|\cdot\|_{D/Y}$. Since $\hat{g}(\infty) = \hat{0}$, $\hat{g}$ admits the following power series expansion

$$\hat{g}(\lambda) = \sum_{k=1}^{\infty} \hat{a}_k \lambda^{-k} \tag{4.7}$$

in a neighborhood of ∞. Since (4.7) converges in the norm $\|\cdot\|_{D/Y}$, we have

$$\|\hat{a}_k\|_{D/Y} \le M^k \quad \text{for some } M > 0 \text{ and } k \in \mathbb{N}.$$

Thus, there are $a_k \in \hat{a}_k$ such that $\|a_k\|_T \le (M+1)^k$, $k \in \mathbb{N}$ and hence the series

$$h(\lambda) = \sum_{k=1}^{\infty} a_k \lambda^{-k}$$

converges in a neighborhood V' of ∞, under the norm of D. Therefore, h is analytic on V' under the norm of D and $\hat{h}(\lambda) = \hat{g}(\lambda)$ on V'. Consequently, $\lambda h(\lambda)$ is analytic on V' and so is $Th(\lambda)$. □

4.16. PROPOSITION. Given T, let $Y \in \text{Inv } T$ with $Y \subset \mathcal{D}_T$ be such that $\hat{T} = T/Y$ is closed. Then, the following properties hold:

(i) If T has the SVEP and $\sigma(T|Y) \cap \sigma(\hat{T})$ is nowhere dense in $\mathbb{C}$, then $Y \in AI(T)$;

(ii) Let $Z \in \text{Inv } T$ be such that $Y \subset Z \subset \mathcal{D}_T$. Then $Z/Y \in AI(\hat{T})$ iff $Z \in AI(T)$.

PROOF. (i): Let $f : \omega_f \to \mathcal{D}_T$ be analytic such that $(\lambda-T)f(\lambda) \in Y$ on an open $\omega_f \subset \mathbb{C}$. We may assume that ω_f is connected. On the quotient space X/Y we have

$$(\lambda-\hat{T})\hat{f}(\lambda) = \hat{0} \quad \text{on } \omega_f.$$

By 4.15, there is an analytic function $h : \omega_h (\subset \omega_f) \to \mathcal{D}_T$ such that $\hat{h}(\lambda) = \hat{f}(\lambda)$ and $(\lambda-T)h(\lambda)$ is analytic on ω_h. Likewise ω_f, ω_h can be assumed to be a connected open set.

First, suppose that $\omega_h \cap \rho(T|Y) \ne \emptyset$. The function $g : \omega_h \cap \rho(T|Y) \to X$, defined by $g(\lambda) = (\lambda-T)h(\lambda)$, is analytic and

$$\hat{g}(\lambda) = (\lambda-\hat{T})\hat{h}(\lambda) = (\lambda-\hat{T})\hat{f}(\lambda) = \hat{0}$$

implies that $g(\lambda) \in Y$ on $\omega_h \cap \rho(T|Y)$. Then

$$(\lambda-T)[h(\lambda) - R(\lambda;T|Y)g(\lambda)] = 0$$

and by the SVEP,

$$h(\lambda) = R(\lambda;T|Y)g(\lambda) \in Y \text{ on } \omega_h \cap \rho(T|Y). \tag{4.8}$$

Thus $h(\lambda) \in Y$ on ω_h, by analytic continuation. Since $\hat{f}(\lambda)$ and $\hat{h}(\lambda)$ agree on ω_h, $f(\lambda) - h(\lambda) \in Y$ on ω_h. In view of (4.8), $f(\lambda) \in Y$ on ω_f, by analytic continuation.

Next, assume that $\omega_h \subset \sigma(T|Y)$. Since, by hypothesis, $\omega_h \cap \rho(\hat{T}) \neq \emptyset$, it follows from $(\lambda-\hat{T})\hat{h}(\lambda) = \hat{0}$ that $\hat{h}(\lambda) = \hat{0}$ on $\omega_h \cap \rho(\hat{T})$. Thus $\hat{f}(\lambda) = \hat{0}$ on ω_f, by analytic continuation and hence $f(\lambda) \in Y$ on ω_f.

(ii). (Only if): Let $f : \omega_f \to \mathcal{D}_T$ be analytic and satisfy condition

$$(\lambda-T)f(\lambda) \in Z \tag{4.9}$$

on an open connected $\omega_f \subset \mathbb{C}$. On the quotient space X/Y, there corresponds

$$(\lambda-\hat{T})\hat{f}(\lambda) \in Z/Y \text{ on } \omega_f.$$

Then, by hypothesis, $\hat{f}(\lambda) \in Z/Y$ on ω_f or, equivalently, $f(\lambda) \in Z$ on ω_f. Thus Z is analytically invariant under T.

(If): Let $\hat{f} : \omega_{\hat{f}} \to X/Y$ be analytic and satisfy condition

$$(\lambda-\hat{T})\hat{f}(\lambda) \in Z/Y \text{ on an open connected } \omega_{\hat{f}} \subset \mathbb{C}.$$

Fix $\lambda_0 \in \omega_{\hat{f}}$. By an argument used in the second part of the proof of 4.14, $\hat{f}$ can be lifted to a $\mathcal{D}_T$-valued function f, analytic on a neighborhood $\omega \subset \omega_{\hat{f}}$ of λ_0 such that $f(\lambda) \in \hat{f}(\lambda)$ on ω. Then (4.9) holds on ω, $f(\lambda) \in Z$ on ω, $\hat{f}(\lambda) \in Z/Y$ on ω and hence $\hat{f}(\lambda) \in Z/Y$ on $\omega_{\hat{f}}$, by analytic continuation. □

4.17. DEFINITION. Given T, $Y \in \text{Inv } T$ is said to be T-*absorbent* if, for any $y \in Y$ and all $\lambda \in \sigma(T|Y)$,

$$(\lambda-T)x = y \tag{4.10}$$

implies that $x \in Y$.

4.18. PROPOSITION. Given T, each T-absorbent space is a ν-space of T.

PROOF. Let Y be a T-absorbent space and suppose that $\sigma(T|Y) \not\subset \sigma(T)$. Then $R(\lambda;T)Y \not\subset Y$ for some $\lambda \in \rho(T) \cap \sigma(T|Y)$ and hence not every solution of (4.10) belongs to Y. This, however, contradicts the definition of Y. □

4.19. PROPOSITION. Given T, let Y be a T-absorbent space. In the following two cases:

(i) $\sigma_p(T) = \emptyset$;

(ii) T has the SVEP, $Y \subset \mathcal{D}_T$ and $\hat{T} = T/Y$ is closed;

Y is analytically invariant under T.

PROOF. Let $f : \omega_f \to \mathcal{D}_T$ be analytic and satisfy condition

$$(\lambda-T)f(\lambda) \in Y \text{ on an open } \omega_f \subset \mathbb{C}.$$

Without loss of generality, we assume that ω_f is connected. If $\omega_f \subset \sigma(T|Y)$ then $f(\lambda) \in Y$, by hypothesis. Therefore, assume that $\omega_f \cap \rho(T|Y) \neq \emptyset$. Since

$$g(\lambda) = (\lambda-T)f(\lambda) \in Y \text{ on } \omega_f \cap \rho(T|Y),$$

we have

$$(\lambda-T)[f(\lambda) - R(\lambda;T|Y)g(\lambda)] = 0 \text{ on } \omega_f \cap \rho(T|Y).$$

In case (i),

$$f(\lambda) = R(\lambda;T|Y)g(\lambda) \in Y \text{ on } \omega_f \cap \rho(T|Y)$$

and hence $f(\lambda) \in Y$ on ω_f, by analytic continuation.

In case (ii), use Lemma 4.15 to assert the existence of a function $h : \omega_h \to \mathcal{D}_T$, analytic on an open connected $\omega_h \subset \omega_f$ such that $\hat{h}(\lambda) = \hat{f}(\lambda)$ and $g(\lambda) = (\lambda-T)h(\lambda)$ is analytic on ω_h. On X/Y, we have

$$\hat{g}(\lambda) = (\lambda-\hat{T})\hat{h}(\lambda) = (\lambda-\hat{T})\hat{f}(\lambda) = \hat{0}$$

and hence $g(\lambda) \in Y$ on ω_h. Since Y is T-absorbent, $h(\lambda) \in Y$ on $\omega_h \cap \sigma(T|Y)$. For $\lambda \in \omega_h \cap \rho(T|Y)$, we have

$$(\lambda-T)[h(\lambda) - R(\lambda;T|Y)g(\lambda)] = 0.$$

$R(\lambda;T|Y)g(\lambda)$ being analytic on ω_h, the SVEP of T implies

$$h(\lambda) = R(\lambda;T|Y)g(\lambda) \in Y \text{ for } \lambda \in \omega_h \cap \rho(T|Y).$$

Thus $h(\lambda) \in Y$ on all of ω_h and hence $\hat{f}(\lambda) = \hat{h}(\lambda) = \hat{0}$ implies that $f(\lambda) \in Y$ on ω_h and $f(\lambda) \in Y$ on ω_f, by analytic continuation. □

4.20. PROPOSITION. Let T have the SVEP and suppose that

$$X = Y_1 + Y_2.$$

If Y_1 and Y_2 are T-absorbent spaces, then

$$\sigma(T|Y_1 \cap Y_2) \subset \sigma(T|Y_1) \cap \sigma(T|Y_2).$$

PROOF. Let $y \in Y_1 \cap Y_2 = Y$ be arbitrary. Then $R(\lambda;T)y \in Y$ on $\rho(T)$, by 4.2 and 4.18. For $\lambda \in \rho(T|Y_1) \cap \rho(T|Y_2) = \rho(T)$, (where the equality follows from 4.5), we have

$$R(\lambda;T|Y_1)y = [R(\lambda;T)|Y_1]y = R(\lambda;T)y \in Y.$$

Y_2 being T-absorbent, for $\lambda \in \rho(T|Y_1) \cap \sigma(T|Y_2)$, $(\lambda-T)R(\lambda;T|Y_1)y = y$ implies that $R(\lambda;T|Y_1)y \in Y_2$. On the other hand, $R(\lambda;T|Y_1)y \in Y_1$ and hence $R(\lambda;T|Y_1)y \in Y$.

Thus, for all $\lambda \in \rho(T|Y_1)$, we have $R(\lambda;T|Y_1)y \in Y$. Now, 4.2 applied to $Y \in \text{Inv } T|Y_1$, gives $\sigma(T|Y) \subset \sigma(T|Y_1)$. By symmetry, $\sigma(T|Y) \subset \sigma(T|Y_2)$ and the assertion of the proposition follows. □

The property expressed by the foregoing theorem can be extended, via induction, to any finite sum decomposition of X into T-absorbent subspaces.

4.21. PROPOSITION. Given T, let $Y \in \text{Inv } T$ be T-absorbent with $\sigma(T|Y)$ compact. Then $\mathcal{T}$, as defined by 3.9, is T-absorbent.

PROOF. Let $y \in \mathcal{T}$, $\lambda \in \sigma(T|\mathcal{T}) = \sigma(T|Y)$, and let x be a solution of

(4.11) $$(\lambda-T)x = y.$$

Y being T-absorbent, $x \in Y$. There is a representation

$$x = x_0 + x_1 \quad \text{with} \quad x_0 \in \mathcal{T},\ x_1 \in W.$$

By (4.11),

$$(\lambda-T)x_0 = y, \quad (\lambda-T)x_1 = 0,$$

and hence $x_1 = 0$, $x = x_0 \in \mathcal{T}$. □

4.22. DEFINITION. Given T, $Y \in \text{Inv } T$ is said to be a *spectral maximal space* of T if, for any $Z \in \text{Inv } T$, the inclusion $\sigma(T|Z) \subset \sigma(T|Y)$ implies $Z \subset Y$.

$Y \in \text{Inv } T$ with $Y \subset \mathcal{D}_T$ is called a T-*bounded spectral maximal space* if conditions $Z \in \text{Inv } T$, $Z \subset \mathcal{D}_T$, $\sigma(T|Z) \subset \sigma(T|Y)$ imply $Z \subset Y$.

We denote by $SM(T)$ and $SM_b(T)$ the family of spectral maximal spaces of T and the family of T-bounded spectral maximal spaces, respectively. Clearly, if $Y \subset \mathcal{D}_T$ is a spectral maximal space of T then Y is a T-bounded spectral maximal space. Conversely, however, not every T-bounded spectral maximal space is a spectral maximal space of T. In fact, if $Y \in SM_b(T)$ and $Z \in \text{Inv } T$ is not contained in $\mathcal{D}_T$, then $\sigma(T|Z) \subset \sigma(T|Y)$ need not imply $Z \subset Y$. For bounded operators, the two concepts coincide.

4.23. PROPOSITION. Given T, every spectral maximal space of T as well as every T-bounded spectral maximal space is hyperinvariant under T.

PROOF. We confine the proof to $Y \in SM(T)$, that of a T-bounded spectral maximal space being similar. Let $A \in B(X)$ commute with T and choose $\lambda \in \mathbb{C}$ such that $|\lambda| > \|A\|$. Then

$$R(\lambda;A) = \sum_{n=0}^{\infty} \lambda^{-n-1}A^n.$$

For every $x \in \mathcal{D}_T$ and $k \in \mathbb{N}$, we have

$$\sum_{n=0}^{k} \lambda^{-n-1}A^nTx = T(\sum_{n=0}^{k} \lambda^{-n-1}A^nx).$$

T being closed, $k \to \infty$ implies that $R(\lambda;A)x \in \mathcal{D}_T$ and

(4.12) $$R(\lambda;A)Tx = TR(\lambda;A)x.$$

Thus $R(\lambda;A)$ commutes with T. Furthermore, the linear manifold $Y_\lambda = R(\lambda;A)Y$ is closed and hence it is a subspace of X. Evidently,

(4.13) $$R(\lambda;A)(Y \cap \mathcal{D}_T) \subset Y_\lambda \cap \mathcal{D}_T.$$

If $y \in Y_\lambda \cap \mathcal{D}_T$, then $(\lambda-A)y \in Y \cap \mathcal{D}_T$ and $R(\lambda;A)(\lambda-A)y = y$. Therefore, (4.13) is an equality

(4.14) $$R(\lambda;A)(Y \cap \mathcal{D}_T) = Y_\lambda \cap \mathcal{D}_T.$$

Then, for $y \in Y_\lambda \cap \mathcal{D}_T$, there is $x \in Y \cap \mathcal{D}_T$, such that $y = R(\lambda;A)x$. Thus (4.12) and (4.14) imply $Ty = TR(\lambda;A)x = R(\lambda;A)Tx \in Y_\lambda$ and hence $Y_\lambda \in \text{Inv } T$. Moreover, it follows from (4.12) and (4.14) that

$$[R(\lambda;A)]^{-1}(T|Y_\lambda)R(\lambda;A)x = (T|Y)x, \quad x \in Y \cap \mathcal{D}_T.$$

Thus $T|Y_\lambda$ and $T|Y$ are similar and hence

(4.15) $$\sigma(T|Y_\lambda) = \sigma(T|Y).$$

Since $Y \in SM(T)$, (4.15) implies that $Y_\lambda \subset Y$. Consequently, for $|\lambda| > \|A\|$, Y is invariant under $R(\lambda;A)$. It follows from

$$A = \lim_{\lambda \to \infty} \lambda[\lambda R(\lambda;A) - I],$$

that Y is invariant under A. □

4.24. PROPOSITION. Given T, every spectral maximal space of T and each T-bounded spectral maximal space is T-absorbent.

PROOF. We confine the proof to a spectral maximal space Y of T. Fix $y_0 \in Y$ and $\lambda_0 \in \sigma(T|Y)$. Suppose, to the contrary, that there is a solution $x_0 \notin Y$ of equation

$$(\lambda_0 - T)x_0 = y_0.$$

The subspace

$$Z = \{z \in X : z = y + \alpha x_0,\ y \in Y,\ \alpha \in \mathbb{C}\}$$

is invariant under T. Let $\mu \in \rho(T|Y)$. For $z \in Z \cap \mathcal{D}_T$, $(\mu - T)z = 0$ implies

$$0 = (\mu-T)y + \alpha(\mu-T)x_0 = [(\mu-T)y + \alpha(\lambda_0-T)x_0] + \alpha(\mu-\lambda_0)x_0 \tag{4.16}$$

and hence $\alpha(\mu-\lambda_0)x_0 \in Y$. Since $\mu \neq \lambda_0$ and $x_0 \notin Y$, it follows that $\alpha = 0$ and then (4.16) implies that $(\mu-T)y = 0$. Since $\mu \in \rho(T|Y)$, $y = 0$ and hence $\mu - T|Z$ is injective.

Next, for $y + \alpha x_0 \in Z$, $\alpha' = \frac{\alpha}{\mu-\lambda_0}$, let $y' \in Y$ be the solution of equation

$$(\mu-T)y' = -\alpha' y_0 + y.$$

Then

$$(\mu-T)(y' + \alpha' x_0) = -\alpha' y_0 + y + \alpha x_0 + \alpha' y_0 = y + \alpha x_0$$

and hence $\mu - T|Z$ is bijective. This implies $\sigma(T|Z) \subset \sigma(T|Y)$ and since $Y \in SM(T)$, $Z \subset Y$. This, however, contradicts the assumption $x_0 \notin Y$. □

Spectral maximality of invariant subspaces, under a bounded operator, can be induced to restrictions and coinduced operators [Ap.1968,c; Proposition 3.2]. Such properties have straightforward extensions to the unbounded case:

4.25. PROPOSITION. Given T, let $Y, Z \in \text{Inv}\, T$ with $Y \subset Z$. Then, the following implications hold:

(i) $Y \in SM(T) \Rightarrow Y \in SM(T|Z)$;

(ii) $Z \in SM(T),\ Y \in SM(T|Z) \Rightarrow Y \in SM(T)$;

(iii) $Y, Z \in SM(T) \Rightarrow Z/Y \in SM(T/Y)$, provided that T/Y is closed;

(iv) $Z \in SM(T),\ Y \in SM_b(T) \Rightarrow Z/Y \in SM(T/Y)$, provided that T/Y is closed.

4.26. COROLLARY. Given T, let $Y \in SM(T)$ or $Y \subset \mathcal{D}_T$ be a T-bounded spectral maximal space. Then, in each of the following cases:

(i) $\sigma_p(T) = \emptyset$;

(ii) T has the SVEP, $Y \subset \mathcal{D}_T$ and T/Y is closed;

Y is analytically invariant under T.

PROOF follows from 4.24 and 4.19. □

4.27. COROLLARY. Given T, if $Y_1, Y_2 \in SM(T)$, (or $Y_1, Y_2 \in SM_b(T)$), then $Y_1 \subset Y_2$ implies $\sigma(T|Y_1) \subset \sigma(T|Y_2)$.

PROOF. Y_1 is T-absorbent and, as invariant under $T|Y_2$, it is $T|Y_2$-absorbent, by 4.24 and 4.25 (i). In particular, Y_1 is a ν-space of $T|Y_2$. □

4.28. COROLLARY. If $T \in B(X)$ has the SVEP then every spectral maximal space of T is analytically invariant under T.

PROOF. For any $Y \in SM(T)$, condition (ii) of 4.26 evidently holds. □

4.29. PROPOSITION. Given T, let $Y \in SM(T)$, (or $Y \in SM_b(T)$). Let $f : \omega_f \to \mathcal{D}_T$ be analytic and satisfy conditions

$$(\lambda-T)f(\lambda) = 0, \quad f(\lambda) \not\equiv 0 \tag{4.17}$$

on an open connected $\omega_f \subset \mathbb{C}$. Then $\omega_f \cap \sigma(T|Y) \neq \emptyset$ implies $\omega_f \subset \sigma(T|Y)$.

PROOF. We confine the proof to $Y \in SM(T)$. Since T is closed, it follows from (4.17) that

$$Tf^{(n+1)}(\lambda) = \lambda f^{(n+1)}(\lambda) + (n+1)f^{(n)}(\lambda), \ \lambda \in \omega_f, \quad n \in Z^+. \tag{4.18}$$

To show that, for each $n \in Z^+$, $f^{(n)}(\lambda) \in Y$ on ω_f, we proceed by induction on n. Let $\lambda_0 \in \omega_f \cap \sigma(T|Y)$. It follows from 4.24 that $(\lambda_0-T)f(\lambda_0) = 0$ implies $f(\lambda_0) \in Y$. Suppose that $f^{(n)}(\lambda_0) \in Y$. Then (4.18) and 4.24 imply that $f^{(n+1)}(\lambda_0) \in Y$. Thus, for each $n \in Z^+$, $f^{(n)}(\lambda_0) \in Y$ and hence $f^{(n)}(\lambda) \in Y$ on some neighborhood of λ_0. Then, for every $n \in Z^+$, $f^{(n)}(\lambda) \in Y$ on ω_f, by analytic continuation.

Next, we show that $\omega_f \subset \sigma(T|Y)$. Since $f(\lambda) \not\equiv 0$, for each $\lambda \in \omega_f$ there is $m \in \mathbb{N}$ such that

$$f(\lambda) = f'(\lambda) = \dots = f^{(m-1)}(\lambda) = 0 \quad \text{and} \quad f^{(m)}(\lambda) \neq 0.$$

Then (4.18) gives

$$Tf^{(m)}(\lambda) = \lambda f^{(m)}(\lambda)$$

and hence $\lambda \in \sigma(T|Y)$. □

4.30. THEOREM. Given T, if $Y \in SM(T)$ with $\sigma(T|Y)$ compact, then $T \in SM_b(T)$.

PROOF. Let $Z \in \text{Inv}\, T$ be such that $Z \subset \mathcal{D}_T$ and $\sigma(T|Z) \subset \sigma(T|T)$. Then, by 3.9, $\sigma(T|Z) \subset \sigma(T|Y)$. Since $Y \in SM(T)$, $Z \subset Y$. For $x \in Z$, $\lambda \in \rho(T|Y)$, $R(\lambda;T|Z)x = R(\lambda;T|Y)x$ implies

$$Qx = \frac{1}{2\pi i} \int_{\partial\Delta} R(\lambda;T|Y)x d\lambda = \frac{1}{2\pi i} \int_{\partial\Delta} R(\lambda;T|Z)x d\lambda = x,$$

where Δ is a bounded Cauchy neighborhood of $\sigma(T|Y)$ and Q is the projection (3.12). Consequently, $x = Qx \in T$ and hence $Z \subset T$. □

4.31. THEOREM. Given T, the following assertions are equivalent:

(I) $\{0\} \in SM(T)$;

(II) for every $Y \in \text{Inv}\, T$ with $\sigma(T|Y)$ compact, $Y \subset \mathcal{D}_T$;

(III) for every $Y \in \text{Inv}\, T$, $Y \neq \{0\}$, implies that $\sigma(T|Y) \neq \emptyset$;

(IV) every T-bounded spectral maximal space is spectral maximal.

PROOF. (I) => (II): For $Y \in \text{Inv}\, T$ with $\sigma(T|Y)$ compact, 3.9 gives the following decomposition of Y:

(4.19) $$Y = T \oplus W, \quad T \subset \mathcal{D}_T, \quad \sigma(T|T) = \sigma(T|Y) \text{ and } \sigma(T|W) = \emptyset.$$

By hypothesis, $W = \{0\}$ and hence $Y = T \subset \mathcal{D}_T$.

(II) => (III): Suppose that $Y \in \text{Inv}\, T$ is such that $Y \neq \{0\}$ and $\sigma(T|Y) = \emptyset$. Since $\sigma(T|Y)$ is compact, $Y \subset \mathcal{D}_T$. Then $T|Y$ is bounded and $\sigma(T|Y) \neq \emptyset$. This, however, contradicts the assumption on $\sigma(T|Y)$.

(III) => (IV): Let $Z \in SM_b(T)$. To see that $Z \in SM(T)$, let $Y \in \text{Inv}\, T$ be such that $\sigma(T|Y) \subset \sigma(T|Z)$. Since $\sigma(T|Z)$ is compact, so is $\sigma(T|Y)$. Then, we have the decomposition (4.19) and, by hypothesis, $\sigma(T|W) = \emptyset$ implies that $W = \{0\}$. Consequently, $Y = T \subset \mathcal{D}_T$ and hence $Y \subset Z$. Therefore, $Z \in SM(T)$.

(IV) => (I): Evidently, $\{0\}$ is a T-bounded spectral maximal space and hence $\{0\}$ is spectral maximal, by (IV). □

The spectral manifold $X(T,\cdot)$ plays a major role in spectral decompositions. It exists if T has the SVEP. If it is closed, it is a subspace of X, invariant (actually, hyperinvariant) under T. Thus, in such a case, we would have a provision of the most important ingredient for the spectral theory: the *invariant subspace*. It is then natural to ask: under what conditions is $X(T,\cdot)$ closed? While the answer has to be deferred until we shall have a powerful machinary of the spectral decomposition at hand, at this time we can explain what kind of a subspace $X(T,\cdot)$ is expected to be.

4.32. THEOREM. Let T have the SVEP. If, for closed $F \subset \mathbb{C}$, $X(T,F)$ is closed then $X(T,F)$ is a spectral maximal space of T and

(4.20) $$\sigma[T|X(T,F)] \subset F \cap \sigma(T).$$

Moreover, in this case, every $Y \in SM(T)$ has a representation

$$Y = X[T,\sigma(T|Y)].$$

PROOF. Let $\lambda \in F^c$ and let $x \in X(T,F)$ be arbitrary. By 2.6 (V),

$$\sigma[x(\lambda),T] = \sigma(x,T) \subset F$$

and hence $x(\lambda) \in X(T,F)$. It follows from $(\lambda-T)x(\lambda) = x$ that $\lambda-T|X(T,F)$ is surjective. Then, 2.5 implies that $\lambda \in \rho[T|X(T,F)]$ and hence

$$\sigma[T|X(T,F)] \subset F.$$

Now (4.20) follows from the fact that $X(T,F)$ is hyperinvariant under T and hence it is a ν-space of T.

Now, let $Y \in \text{Inv } T$ be such that $\sigma(T|Y) \subset \sigma[T|X(T,F)]$. If $x \in Y$ then

$$\sigma(x,T) \subset \sigma(T|Y) \subset \sigma[T|X(T,F)] \subset F$$

and hence $x \in X(T,F)$. Therefore, $Y \subset X(T,F)$ implying that $X(T,F) \in SM(T)$.

Finally, suppose that $Y \in SM(T)$. Since, for every $x \in Y$, $\sigma(x,T) \subset \sigma(T|Y)$, one has

(4.21) $$Y \subset X[T,\sigma(T|Y)].$$

On the other hand, for $F = \sigma(T|Y)$, (4.20) implies

$$\sigma\{T|X[T,\sigma(T|Y)]\} \subset \sigma(T|Y).$$

Since $Y \in SM(T)$, the opposite of inclusion (4.21) follows. □

The hypotheses of the foregoing theorem will play an important role in the spectral decomposition problem. Later, they will be referred to as *property* (κ).

4.33. LEMMA. Let T have the SVEP and suppose that $Y \in \text{Inv } T$ is such that $Y \subset \mathcal{D}_T$. Then $x \in Y$ and $\sigma(x,T) = \emptyset$ imply that $x = 0$.

PROOF. It follows from $\sigma(x,T) = \emptyset$ that the local resolvent is an entire function. If $x \in Y$, then

$$x(\lambda) = R(\lambda;T|Y)x \quad \text{for} \quad |\lambda| > \|T|Y\|.$$

Hence, for $\Gamma = \{\lambda : |\lambda| = \|T|Y\| + 1\}$, one obtains

$$x = \frac{1}{2\pi i}\int_\Gamma R(\lambda;T|Y)x\,d\lambda = \frac{1}{2\pi i}\int_\Gamma x(\lambda)\,d\lambda = 0. \quad \square$$

4.34. THEOREM. Let T have the SVEP and suppose that, for F compact, $X(T,F)$ is closed. There exists a T-bounded spectral maximal space $\Xi(T,F)$ with the following properties

(I) $\quad X(T,F) = \Xi(T,F) \oplus X(T,\emptyset)$;

(II) $\quad \sigma[T|\Xi(T,F)] = \sigma[T|X(T,F)]$.

PROOF. Since $\sigma[T|X(T,F)] \subset F$ is compact, for $Y = X(T,F)$ and $T = \Xi(T,F)$, 3.9 implies (II) and

$$X(T,F) = \Xi(T,F) \oplus W \quad \text{with} \quad \sigma(T|W) = \emptyset.$$

By hypothesis and 4.32, $X(T,F)$ and $X(T,\emptyset)$ are spectral maximal spaces of T. Then, it follows from 4.30 that $\Xi(T,F)$ is a T-bounded spectral maximal space. Consequently,

$$\sigma(T|W) = \emptyset = \sigma[T|X(T,\emptyset)]$$

implies that $W \subset X(T,\emptyset)$. To prove the opposite inclusion, let $x \in X(T,\emptyset)$. Then $x \in X(T,F)$ and hence $Qx \in \Xi(T,F)$, where Q is the projection (3.12). The local resolvent $x(\lambda) \in X(T,F)$ of $x \in X(T,\emptyset)$, is a $\mathcal{D}_T$-valued entire function and $Qx(\lambda) \in \Xi(T,F)$. It follows from

$$(\lambda-T)Qx(\lambda) = Q(\lambda-T)x(\lambda) = Qx$$

and 4.33 that $Qx = 0$. $\square$

For compact $F \subset \mathbb{C}$, $\Xi(T,F)$ will be referred to as the T-*bounded spectral maximal space associated to* $X(T,F)$.

4.35. PROPOSITION. Given T with the SVEP, let $Y \in AI(T)$ be such that $Y \subset \mathcal{D}_T$ and $\hat{T} = T/Y$ is closed. Then, for $x \in X$ and $\hat{x} \in X/Y$,

$$\sigma(x,T) = [\sigma(x,T) \cap \sigma(T|Y)] \cup \sigma(\hat{x},\hat{T}).$$

PROOF. $\hat{T}$ has the SVEP by 4.14. For each $x \in X$, we clearly have $\rho(x,T) \subset \rho(\hat{x},\hat{T})$ and hence

$$\sigma(x,T) \supset [\sigma(x,T) \cap \sigma(T|Y)] \cup \sigma(\hat{x},\hat{T}).$$

To prove the proposition, we shall obtain

$$\rho(x,T) \supset [\rho(x,T) \cup \rho(T|Y)] \cap \rho(\hat{x},\hat{T}).$$

Clearly, it suffices to show that

$$\rho(x,T) \supset \rho(T|Y) \cap \rho(\hat{x},\hat{T}). \tag{4.22}$$

Let $\lambda_0 \in \rho(T|Y) \cap \rho(\hat{x},\hat{T})$ and let $\hat{x}(\cdot)$ be the local resolvent of $\hat{x}$. We have

$$(\lambda-\hat{T})\hat{x}(\lambda) = \hat{x}$$

on a neighborhood $\delta \subset \rho(T|Y) \cap \rho(\hat{x},\hat{T})$ of λ_0. It follows from 4.15 that there exists a function $h : \omega_h \to \mathcal{D}_T$ with both h and Th analytic on a neighborhood $\omega_h \subset \delta$ of λ_0 such that $\hat{h}(\lambda) = \hat{x}(\lambda)$ if $\lambda \in \omega_h$. Put

$$g(\lambda) = (\lambda-T)h(\lambda) - x.$$

Then

$$(\lambda-T)[h(\lambda) - R(\lambda;T|Y)g(\lambda)] = x$$

implies that $\lambda_0 \in \rho(x,T)$ and hence (4.22) follows. □

4.36. COROLLARY. Given $T \in B(X)$ with the SVEP, let $Y \in SM(T)$. Then, for $x \in X$, $\hat{x} \in X/Y$ and $\hat{T} = T/Y$, we have

$$\sigma(x,T) = [\sigma(x,T) \cap \sigma(T|Y)] \cup \sigma(\hat{x},\hat{T}).$$

PROOF. Y being analytically invariant under T, by 4.28, the assertion of the corollary follows directly from 4.35. □

NOTES AND COMMENTS.

The single valued extension property, introduced by N. Dunford [Du.1952,1954,1958], appears as condition (A) among the necessary and sufficient conditions for a bounded operator to be *spectral*. Extensions to the unbounded case are due to W.G. Bade [Ba.1954], J. Schwartz [S.1954] and others. In an early stage of the theory of the Dunford-type spectral operators, it was shown that not every operator has the SVEP. A counter-example appeared in [C-Fo.1968, Example 1.7]. J.K. Finch [F.1975] gave some necessary and sufficient conditions for a closed operator to possess the SVEP. His proving techniques were expanded in the proof of 2.2, which together with 2.3 appeared in [W-E]. Corollary 2.4, as well as 2.5, appears explicitly in [F.1975]. Corollary 2.3 was independently proved by

Zou, Cheng Zu, for bounded operators. Proposition 2.6 contains known properties of the local spectrum. For bounded operators, (I), (II), (III) and (VI) appeared in [Du.1952], [Du.1954] and (V) in [C-Fo.1967], [C-Fo.1968]. Proposition 2.7 was proved in [Si.1964]. The spectral manifold concept makes its appearance in Dunford's condition (C), [Du.1958], [Bar.1964] and its closure was employed in [Bi.1959], under the name of *strong spectral manifold*. Later, it was fully developed in [Fo.1963] for the theory of decomposable operators.

Some results on the stability of the SVEP under functional calculus are to be found in [C-Fo.1967]. Apostol [Ap.1968,c] and independently, Bartle and Kariotis [Bar-Ka.1973] obtained a local version of the spectral mapping theorem. Its extension to the unbounded case is in [V.1982]. Part of 3.1 was proved in [B.1976], for bounded operators. Its extension to to the unbounded case is in [W-E.1984]. Also 3.2 appeared in the latter. Proposition 3.5 is in [W-E]. A bounded version of 3.6 appeared in [Sc.1959]. Corollary 3.9 is in [E-W.1984,a] and 3.11, with many applications, is part of a theorem in [Fo.1968].

The concepts of the ν- and μ-spaces, with many useful properties, are due to Bartle and Kariotis [Bar-Ka.1973].In this vein, we mention 4.3, 4.8 and 4.13 [ibid.]. Proposition 4.2 is in [Sc.1959]. The proof of 4.6 appeared in [Do.1969] and [Do.1978, Lemma 19.1]. Proposition 4.9 and the bounded version of 4.25 are included in [Ap.1968,c]. The analytically invariant subspace was introduced by Frunză [Fr.1973] and he is the author of 4.14, confined to a bounded operator [ibid.]. Also, 4.36 appeared in [Fr.1973]. Lemma 4.15 appeared in [N.1981], as indicated in the text. The T-absorbent space originates in a work of Vasilescu [V.1969]. The spectral maximal space concept for bounded operators, with many spectral properties, referred to as 4.23, 4.29 and 4.32, appeared in [Fo.1963]. Proposition 4.29 appeared in [E.1980,a]. Theorem 4.31 originates from [Wa.1981] and, with some added properties, appeared in [E-W.1984,a]. Proposition 4.35 (in its bounded version, 4.36) appeared in [Fr.1973].

Finally, it is worth mentioning that the analytically invariant subspace concept was generalized under the name of *strongly analytic subspace* and subsequently gave rise to interesting properties and useful characterizations within the spectral decomposition problem [L.1979] and [Sn.1984].

CHAPTER II. THE SPECTRAL DECOMPOSITION PROPERTY

Now that we have described the invariant subspaces we shall work with, we are prepared to study the spectral decomposition problem.

§.5. THE 1-SPECTRAL DECOMPOSITION PROPERTY.

We begin our study with the 2-summand spectral decomposition of the underlying Banach space X by a closed operator T. After many of the desired results are obtained, it will come as a pleasant surprise that this particular problem is equivalent to the general spectral decomposition problem.

As mentioned in the Introduction, we assume throughout this work that every open cover of a set $S \subset \mathbb{C}$ has, at most, one unbounded member. For $\{G_i\}_{i=0}^n \in \text{cov } S$ with $G_0 \in V_\infty$, the sets G_i $(1 \le i \le n)$ are assumed to be relatively compact, in symbols, $\{G_i\}_{i=1}^n \subset G^K$.

5.1. DEFINITION. Given $n \in \mathbb{N}$, T is said to have the *n-spectral decomposition property* (n-SDP) if, for every $\{G_i\}_{i=0}^n \in \text{cov } \sigma(T)$ with $G_0 \in V_\infty$ and $\{G_i\}_{i=1}^n \subset G^K$, there exists $\{X_i\}_{i=0}^n \subset \text{Inv } T$, satisfying the following conditions:

(I) $X_i \subset \mathcal{D}_T$ if G_i $(1 \le i \le n)$ is relatively compact;

(II) $\sigma(T|X_i) \subset G_i$, $0 \le i \le n$;

(III) $X = \sum_{i=0}^{n} X_i$.

If, for every $n \in \mathbb{N}$, T has the n-SDP, then T is said to have the *spectral decomposition property* (SDP).

5.2. REMARKS. Without any deviation from the notion introduced in 5.1, we may

(a) consider $\{G_i\}_{i=0}^n \in \text{cov } \mathbb{C}$ (see Notes and Comments);

(b) substitute (II) by

(II') $\sigma(T|X_i) \subset \overline{G}_i$, $0 \le i \le n$.

5.3. EXAMPLE. If $\sigma(T)$ is totally disconnected, then T has the 1-SDP.

PROOF. Let $\{G_0, G_1\} \in \text{cov } \sigma(T)$ with $G_0 \in V_\infty$ and $G_1 \in G^K$. By the hypothesis on $\sigma(T)$, there is $\{H_0,H_1\} \in \text{cov}\,\sigma(T)$ with $\overline{H}_i \subset G_i$ $(i=0,1)$, $H_0 \in V_\infty$ and $\overline{H}_0 \cap \overline{H}_1 = \emptyset$. The disjoint spectral sets $\sigma_i = \overline{H}_i \cap \sigma(T)$, $(i=0,1)$ with σ_1 compact, are such that

$$\sigma(T) = \sigma_0 \cup \sigma_1 .$$

By the functional calculus, X admits a direct sum decomposition

$$X = X_0 \oplus X_1,$$

with

$$\sigma(T|X_i) \subset \sigma_i \subset G_i, \quad i=0,1.$$

Then $X_1 \subset \mathcal{D}_T$, $T|X_1$ is bounded and hence T has the 1-SDP. □

5.4. DEFINITION. An operator T is said to have *property* (κ) if

(a) T has the SVEP; (b) $X(T,F)$ is closed for every closed F.

An equivalent version of Bishop's condition (β), [Bi.1959, Definition 8], can be stated in terms of the following

5.5. DEFINITION. T has *property* (β) if, for any sequence $\{f_n : G \to \mathcal{D}_T\}$ of analytic functions, $(\lambda-T)f_n(\lambda) \to 0$ (as $n\to\infty$), in the strong topology of X and uniformly on every compact subset of G, implies that $f_n(\lambda) \to 0$ in the strong topology of X and uniformly on every compact subset of G.

5.6. PROPOSITION. For every closed T, property (β) implies property (κ).

PROOF. Assume that T has property (β). Clearly, T has the SVEP. Let F be closed and let a sequence $\{x_n\} \subset X(T,F)$ converge to a point $x \in X$. Since, for every n, $\rho(x_n,T) \supset F^c = G$, there exists a sequence $\{f_n : G \to \mathcal{D}_T\}$ of analytic functions, such that

$$(\lambda-T)f_n(\lambda) = x_n \text{ for all } \lambda \in G,\ n \in \mathbb{N}.$$

Next, we show that property (β) implies that $\{f_n\}$ is a Cauchy sequence on every compact subset of G. Otherwise, there is a compact $K \subset G$, a positive number ε and there are subsequences $\{m_i\}$ and $\{n_i\}$ of $\{n\}$ such that, for each i, there exists $\lambda_i \in K$ satisfying

$$\|f_{m_i}(\lambda_i) - f_{n_i}(\lambda_i)\| \geq \varepsilon .$$

Property (β) applied to the sequence $\{f_{m_i} - f_{n_i}\}_i$ implies that $f_{m_i} - f_{n_i} \to 0$ (as $i\to\infty$), uniformly on every compact subset of G and, in particular, on K. But this contradicts the above inequality.

Consequently, $\{f_n\}$ converges to a function f, uniformly on every compact subset of G. Therefore, $f : G \to \mathcal{D}_T$ is analytic and since T is closed,

$$(\lambda - T)f(\lambda) = x \quad \text{on } G.$$

Thus $\lambda \in \rho(x,T)$, i.e. $\sigma(x,T) \subset F$ and hence $x \in X(T,F)$, which proves that X(T,F) is closed. □

5.7. LEMMA. Given a subspace Y of X, let H and K be open disks with $\overline{K} \subset H$. If $\hat{f} : V \to X/Y$ is analytic on a neighborhood V of $\overline{H}$, then there exists an analytic function $h : H \to X$ and a constant A, independent of $\hat{f}$, such that

$$\max_{\lambda \in \overline{K}} \|h(\lambda)\| \le A \max_{\lambda \in \overline{H}} \|\hat{f}(\lambda)\|.$$

PROOF. For $\lambda_0 \in \mathbb{C}$, $\lambda_1 \in H$, $0 < r \le r + |\lambda_0 - \lambda_1| < R$, define

$$H = \{\lambda : |\lambda - \lambda_0| < R\}, \quad K = \{\lambda : |\lambda - \lambda_1| < r\}$$

and let

$$\hat{f}(\lambda) = \sum_{n=0}^{\infty} \hat{a}_n (\lambda - \lambda_0)^n \quad \text{with} \quad \{\hat{a}_n\} \subset X/Y$$

be the power series expansion of $\hat{f}$. Choose ρ to satisfy

$$r + |\lambda_0 - \lambda_1| < \rho < R.$$

By the Cauchy inequality, we have

$$\|\hat{a}_n\| \le MR^{-n}, \quad \text{where} \quad M = \max_{\lambda \in \overline{H}} \|\hat{f}(\lambda)\|.$$

For every n, choose $a_n \in \hat{a}_n$ such that $\|a_n\| \le 2\,\|\hat{a}_n\|$ and define

$$h(\lambda) = \sum_{n=0}^{\infty} a_n (\lambda - \lambda_0)^n.$$

Then h is analytic on H and since $\overline{K} \subset \{\lambda : |\lambda - \lambda_0| < \rho\}$, we have

$$\max_{\lambda \in \overline{K}} \|h(\lambda)\| \le \sum_{n=0}^{\infty} \|a_n\| \rho^n < 2M \sum_{n=0}^{\infty} \left(\frac{\rho}{R}\right)^n = A \max_{\lambda \in \overline{H}} \|\hat{f}(\lambda)\|,$$

where $A = \dfrac{2R}{R - \rho}$. □

5.8. THEOREM. Given T, suppose that, for every pair of open disks G, H with $\overline{G} \subset H$, there exists $Y \in \text{Inv } T$ satisfying the following conditions

(a) $\sigma(T|Y) \subset G^c$;

(b) $\hat{T} = T/Y$ is bounded and $\sigma(\hat{T}) \subset H$.

Then T has property (β).

PROOF. Let $\{f_n\}$ be a sequence of $\mathcal{D}_T$-valued analytic functions on an open $G_0 \subset \mathbb{C}$, such that

(5.1) $$(\lambda - T)f_n(\lambda) \to 0 \quad (\text{as } n \to \infty)$$

in the strong topology of X and uniformly on every compact subset of G_0. We may assume that $G_0 = \{\lambda : |\lambda| < R\}$ for some $R > 0$. Choose numbers R_0, R_1, R_2 such that $0 < R_0 < R_1 < R_2 < R$ and let $K = \{\lambda : |\lambda| \le R_0\}$, $G = \{\lambda : |\lambda| < R_1\}$, $H = \{\lambda : |\lambda| < R_2\}$. By hypothesis, there exists $Y \in \text{Inv } T$ satisfying conditions (a) and (b) for G and H. It follows from (5.1) that

$$(\lambda - \hat{T})\hat{f}_n(\lambda) \to \hat{0}$$

in the strong topology of X/Y and uniformly on $\overline{H}$. since $\partial H \subset \rho(\hat{T})$, we have

(5.2) $$\hat{f}_n(\lambda) \to \hat{0}$$

in the strong topology of X/Y and uniformly on ∂H. By the maximum principle, the convergence (5.2) is uniform on $\overline{H}$. Futhermore, since $\hat{T}$ is bounded,

(5.3) $$\hat{T}\hat{f}_n(\lambda) \to \hat{0}$$

in the strong topology of X/Y and uniformly on $\overline{H}$.

The graph $\mathcal{G}(T)$ of T is closed in $X \oplus X$ and $\mathcal{G}(T|Y)$ is closed in $Y \oplus Y$. The map

$$\tau : [x \oplus Tx + \mathcal{G}(T|Y)] \to (x + Y) \oplus (Tx + Y)$$

of $\mathcal{G}(T)/\mathcal{G}(T|Y)$ into $\mathcal{G}(\hat{T}) \subset (X/Y) \oplus (X/Y)$ is bijective. Since $\hat{T}$ is bounded, $\mathcal{G}(\hat{T})$ is closed and hence it follows from the inequalities

$$\|x \oplus Tx + \mathcal{G}(T|Y)\| = \inf \{ \|x \oplus Tx + y \oplus Ty\| : y \in Y \cap \mathcal{D}_T\}$$
$$\ge \inf \{ \|(x+y_1) \oplus (Tx+y_2)\| : y_1, y_2 \in Y\} = \|(x+Y) \oplus (Tx+Y)\|,$$

that τ is a topological isomorphism.

Since $\lambda \to \hat{f}_n(\lambda) \oplus \hat{\hat{T}}\hat{f}_n(\lambda)$ is analytic on a neighborhood of $\overline{H}$, so is $\lambda \to \tau^{-1}[\hat{f}_n(\lambda) \oplus \hat{\hat{T}}\hat{f}_n(\lambda)]$. Evidently, $\tau^{-1}(\hat{f}_n \oplus \hat{\hat{T}}\hat{f}_n)$ is a $G(T)/G(T/Y)$ -valued function. Consequently, one can find a $G(T)$ -valued analytic function $h_n \oplus Th_n$, defined on H such that

$$h_n(\lambda) \oplus Th_n(\lambda) \in \tau^{-1}[\hat{f}_n(\lambda) \oplus \hat{\hat{T}}\hat{f}_n(\lambda)].$$

It follows from 5.7 that one can choose h_n such that

$$\max_{\lambda \in K} \| h_n(\lambda) \oplus Th_n(\lambda) \| \leq A \max_{\lambda \in \overline{H}} \| \tau^{-1}[\hat{f}_n(\lambda) \oplus \hat{\hat{T}}\hat{f}_n(\lambda)] \| \tag{5.4}$$

for some $A > 0$. Clearly, both h_n, Th_n are analytic and $\hat{h}_n(\lambda) = \hat{f}_n(\lambda)$ on H. Thus, $h_n(\lambda) - f_n(\lambda) \in Y$ on H. In view of (5.2) and (5.3),

$$\hat{f}_n(\lambda) \oplus \hat{\hat{T}}\hat{f}_n(\lambda) \to \hat{0} \tag{5.5}$$

uniformly on $\overline{H}$. By (5.4) and (5.5),

$$h_n(\lambda) \to 0 \quad \text{and} \quad Th_n(\lambda) \to 0 \tag{5.6}$$

uniformly on K. It follows from (5.1) and (5.6) that

$$(\lambda - T)[h_n(\lambda) - f_n(\lambda)] \to 0 \tag{5.7}$$

uniformly on K. Since $K \subset G$ and $\sigma(T|Y) \subset G^c$, we have $K \subset \rho(T|Y)$ and then (5.7) implies that $h_n(\lambda) - f_n(\lambda) \to 0$ uniformly on K. Thus, by (5.6), $f_n(\lambda) \to 0$ uniformly on K. Since $R_0 < R$ is arbitrary, it follows that T has property (β).

Observe that the inclusion in condition (b) can be substituted by $\sigma(\hat{T}) \subset \overline{G}$. □

5.9. COROLLARY. If T has the 1-SDP then T has property (β).

PROOF. Let G, H be open disks with $\overline{G} \subset H$. Since $\{(\overline{G})^c, H\} \in \operatorname{cov} \sigma(T)$, there are $Y, Z \in \operatorname{Inv} T$ such that

$$X = Y + Z, \quad \sigma(T|Y) \subset (\overline{G})^c \subset G^c, \quad Z \subset \mathcal{D}_T, \quad \sigma(T|Z) \subset H.$$

Since $\rho(T|Y \cap Z) \supset \rho_\infty(T|Z) \supset H^c$, we have $\sigma(T|Y \cap Z) \subset H$. The operators $\hat{T} = T/Y$ and $\tilde{T} = (T|Z)/Y \cap Z$ being similar, by 3.4, $\hat{T}$ is bounded and

$$\sigma(\hat{T}) = \sigma(\tilde{T}) \subset \sigma(T|Z) \cup \sigma(T|Y \cap Z) \subset H.$$

Thus, the hypotheses of 5.8 are satisfied and hence T has property (β). □

5.10. COROLLARY. Let T have the 1-SDP. For every $\{G_0,G_1\} \in \operatorname{cov}\sigma(T)$ with $G_0 \in V_\infty$ and $G_1 \in G^K$, we have

(5.8) $$X = X(T,\overline{G}_0) + \Xi(T,\overline{G}_1).$$

PROOF. T has property (κ), by 5.9 and 5.6. There are $X_0, X_1 \in \operatorname{Inv} T$ with

$$X = X_0 + X_1,\quad X_1 \subset \mathcal{D}_T,\quad \sigma(T|X_i) \subset G_i,\quad i=0,1.$$

Since, for $x_i \in X_i$, $\sigma(x_i,T) \subset \sigma(T|X_i) \subset G_i$ implies $x_i \in X(T,\overline{G}_i)$, it follows that $X_i \subset X(T,\overline{G}_i)$, i=0,1. Consequently, one can write

(5.9) $$X = X(T,\overline{G}_0) + X(T,\overline{G}_1).$$

For the second summand of (5.9), 4.34 gives

(5.10) $$X(T,\overline{G}_1) = \Xi(T,\overline{G}_1) \oplus X(T,\emptyset).$$

Since $X(T,\emptyset) \subset X(T,\overline{G}_0)$, (5.8) follows from (5.9) and (5.10). □

In view of 4.32 and 4.34, (5.8) provides a spectral decomposition of X by T into a linear sum of a T-bounded spectral maximal space and a spectral maximal space of T. Conversely, if T has property (κ) and $\{G_0,G_1\} \in \operatorname{cov}\sigma(T)$, with $G_0 \in V_\infty$ and $G_1 \in G^K$, produces the decomposition (5.8), then T has the 1-SDP. In fact, $\Xi(T,\overline{G}_1) \subset \mathcal{D}_T$,

$$\sigma[T|X(T,\overline{G}_0)] \subset \overline{G}_0,\quad \sigma[T|\Xi(T,\overline{G}_1)] \subset \overline{G}_1$$

fulfill the defining conditions of 5.1, for $n=1$. Next, we investigate for some further properties of the T-bounded spectral maximal spaces of a closed T with the 1-SDP.

5.11. LEMMA. Given T with the 1-SDP, let $F \subset \mathbb{C}$ be compact. Then $x \in \Xi(T,F)$ iff

$$\text{(i)}\ \sigma(x,T) \subset F \quad\text{and}\quad \text{(ii)}\ \lim_{\lambda\to\infty} x(\lambda) = 0.$$

PROOF. (Only if): Let $x \in \Xi(T,F)$. Then

$$\sigma(x,T) \subset \sigma[T|\Xi(T,F)] = \sigma[T|X(T,F)] \subset F.$$

Furthermore, $T|\Xi(T,F)$ being bounded,

$$\lim_{\lambda\to\infty} x(\lambda) = \lim_{\lambda\to\infty} R[\lambda;T|\Xi(T,F)]x = 0.$$

(If): By (i), $x \in X(T,F)$. Since T is closed, it follows from (ii) and from $\lambda x(\lambda) - x = Tx(\lambda)$ that

$$\lim_{\lambda\to\infty} [\lambda x(\lambda) - x] = T \lim_{\lambda\to\infty} x(\lambda) = 0.$$

The function $f : V \to X$, defined by $f(\lambda) = \lambda x(\lambda) - x$ is analytic on a neighborhood V of ∞ and $f(\infty) = \lim_{\lambda\to\infty} f(\lambda) = 0$. Let $r > 0$ be sufficiently large for

$$F \subset \{\lambda : |\lambda| < r\} \text{ and } V \supset \{\lambda : |\lambda| \geq r\}.$$

Note that ∞ is, at least, a double zero of $\frac{f(\lambda)}{\lambda} = x(\lambda) - \frac{x}{\lambda}$. Consequently,

$$x = \frac{1}{2\pi i}\int_\Gamma \frac{x}{\lambda}d\lambda = \frac{1}{2\pi i}\int_\Gamma x(\lambda)d\lambda - \frac{1}{2\pi i}\int_\Gamma \frac{f(\lambda)}{\lambda}d\lambda$$

$$= \frac{1}{2\pi i}\int_\Gamma R[\lambda;T|X(T,F)]x d\lambda = Qx \in \Xi(T,F),$$

where $\Gamma = \{\lambda : |\lambda| = r\}$ and $Q = \frac{1}{2\pi i}\int_\Gamma R[\lambda;T|X(T,F)]d\lambda$. □

The foregoing lemma extends the *monotonic property* of $X(T,\cdot)$ to T-bounded spectral maximal spaces: if F_1 and F_2 are compact, then $F_1 \subset F_2$ implies $\Xi(T,F_1) \subset \Xi(T,F_2)$.

5.12. THEOREM. Given T with the 1-SDP, let F_1 be a closed and F_2 a compact subset of $\mathbb{C}$. If F_1 and F_2 are disjoint, then

$$X(T,F_1 \cup F_2) = X(T,F_1) \oplus \Xi(T,F_2).$$

PROOF. By denoting $F = F_1 \cup F_2$, one obtains easily

(5.11) $$X(T,F) \supset X(T,F_1) + \Xi(T,F_2).$$

On the other hand, by the functional calculus, $X(T,F)$ admits a direct sum decomposition

$$X(T,F) = X_1 \oplus X_2,$$

where $X_1, X_2 \in \text{Inv}\, T$, $\sigma(T|X_i) \subset F_i$ $(i=1,2)$ and $X_2 \subset \mathcal{D}_T$. Then $X_i \subset X(T,F_i)$, $i=1,2$. Since $T|X_2$ is bounded, for every $x \in X_2$, one has $\sigma(x,T) \subset F_2$ and then

$$\lim_{\lambda\to\infty} x(\lambda) = \lim_{\lambda\to\infty} R(\lambda;T|X_2)x = 0.$$

Thus, $x \in \Xi(T,F_2)$, by 5.11, and hence $X_2 \subset \Xi(T,F_2)$. Consequently, the opposite of (5.11) holds and hence

(5.12) $$X(T,F) = X(T,F_1) + \Xi(T,F_2).$$

To see that (5.12) is a direct sum, suppose that

$$x \in X(T,F_1) \cap \Xi(T,F_2) \subset X(T,F_1) \cap X(T,F_2) = X(T,\emptyset).$$

Then $\sigma(x,T) = \emptyset$ and since $x \in \Xi(T,F_2) \subset \mathcal{D}_T$, $x = 0$, by 4.33. □

5.13. THEOREM. Given T with the 1-SDP, let F_1 be closed and F_2 compact. Then

$$\Xi(T,F_1 \cap F_2) = X(T,F_1) \cap \Xi(T,F_2). \tag{5.13}$$

PROOF. The inclusion

$$\Xi(T,F_1 \cap F_2) \subset X(T,F_1) \cap \Xi(T,F_2) \tag{5.14}$$

is immediate from 5.11. Conversely, let $x \in X(T,F_1) \cap \Xi(T,F_2)$. Then $\sigma(x,T) \subset F_1 \cap F_2$ and therefore $x \in X(T,F_1 \cap F_2)$. Since $x \in \Xi(T,F_2)$, 5.11 implies $\lim_{\lambda\to\infty} x(\lambda) = 0$. Quote again 5.11 and infer that $x \in \Xi(T,F_1 \cap F_2)$. Thus, the opposite of (5.14) follows and hence (5.13) holds. □

5.14. THEOREM. Given T with the 1-SDP, for every open G,

$$\overline{G \cap \sigma(T)} \subset \sigma[T|X(T,\overline{G})] \subset \overline{G} \cap \sigma(T).$$

PROOF. Let $\lambda \in G \cap \sigma(T)$. Choose $\{G_0,G_1\} \in \operatorname{cov} \sigma(T)$ with $G_0 \in V_\infty$, $G_1 \in G^K$, $\lambda \notin \overline{G}_0$ and $\lambda \in G_1 \subset G$. By the 1-SDP,

$$X = X(T,\overline{G}_0) + \Xi(T,\overline{G}_1) = X(T,\overline{G}_0) + X(T,\overline{G}_1) = X(T,\overline{G}_0) + X(T,\overline{G}).$$

Consequently,

$$\sigma(T) \subset \sigma[T|X(T,\overline{G}_0)] \cup \sigma[T|X(T,\overline{G})].$$

Since $\lambda \notin \overline{G}_0$ and $\sigma[T|X(T,\overline{G}_0)] \subset \overline{G}_0$, it follows that $\lambda \in \sigma[T|X(T,\overline{G})]$ and hence $\overline{G \cap \sigma(T)} \subset \sigma[T|X(T,\overline{G})]$. Now 4.32 concludes the proof. □

5.15. LEMMA. Given a *linear operator* $T : \mathcal{D}_T(\subset X) \to X$, let $X_0,X_1 \in \operatorname{Inv} T$ satisfy conditions

$$X = X_0 + X_1, \quad X_1 \subset \mathcal{D}_T, \quad T|X_1 \in B(X_1). \tag{5.15}$$

Then T is closed iff $T|X_0$ is closed and T is densely defined iff $T|X_0$ is densely defined. In particular, if T is closed, densely defined and has the 1-SDP then, for closed $F \in V_\infty$, $T|X(T,F)$ is densely defined in $X(T,F)$.

PROOF. Evidently, if T is closed so is $T|X_0$. Suppose that $T|X_0$ is closed and let $\{y_n\} \subset \mathcal{D}_T$ be such that

$$\text{(a)}\quad y_n \to y; \quad \text{(b)}\quad Ty_n \to z. \tag{5.16}$$

In view of (5.15) and 3.10, there is $M > 0$ such that each y_n has a representation

$$y_n = y_{n0} + y_{n1}, \quad y_{ni} \in X_i \quad (i=0,1)$$

satisfying condition

(5.17) $$\|y_{n+1,0} - y_{n0}\| + \|y_{n+1,1} - y_{n1}\| \le M\|y_{n+1} - y_n\|.$$

Without loss of generality, we may assume that $\sum_{n=1}^{\infty} \|y_{n+1} - y_n\| < \infty$. Then (5.17) implies

(5.18) $$y_{ni} \to y_i, \quad y_i \in X_i \quad (i=0,1).$$

$T|X_1$ being bounded, it follows from (5.18) that $Ty_{n1} \to Ty_1$ and hence

(5.19) $$Ty_{n0} \to z - Ty_1,$$

by (5.16,b). Since $T|X_0$ is closed, (5.18) and (5.19) imply that

$$y_0 \in X_0 \cap \mathcal{D}_T \quad \text{and} \quad Ty_0 = z - Ty_1.$$

Therefore, $y = y_0 + y_1 \in \mathcal{D}_T$ and $Ty = T(y_0 + y_1) = z$ and hence T is closed.

Next, suppose that T is densely defined. Let $x \in X_0$. There is a sequence $\{x_n\} \subset \mathcal{D}_T$ converging to x. There is $M > 0$ and, for every n, there is a representation

$$x - x_n = x_{n0} + x_{n1}, \quad x_{ni} \in X_i \quad (i=0,1)$$

with

$$\|x_{n0}\| + \|x_{n1}\| \le M\|x - x_n\|.$$

Then $x_{ni} \to 0$ $(i=0,1)$. Since, by hypothesis,

$$y_n = x - x_{n0} = x_n + x_{n1} \in \mathcal{D}_T \cap X_0,$$

it follows that

$$\|x - y_n\| = \|x_{n0}\| \to 0.$$

Thus $T|X_0$ is densely defined. Conversely, if $T|X_0$ is densely defined, then it follows from $X_1 \subset \mathcal{D}_T$ that T is densely defined.

Finally, assume that T is closed, densely defined and has the 1-SDP. If F is a closed neighborhood of ∞, then for any $G \in G^K$ with $\{\text{Int } F, G\} \in \text{cov } \sigma(T)$, (5.15) holds in terms of $X_0 = X(T,F)$ and $X_1 = \Xi(T,\overline{G})$. □

5.16. THEOREM. Let T have the 1-SDP. Then, for every $Y \in \text{Inv } T$ with $\sigma(T|Y) \neq \mathbb{C}$, $\hat{T} = T/Y$ is closed. In particular, if $Y \in SM(T)$ or $Y \in SM_b(T)$, then

(5.20) $$\sigma(\hat{T}) = \overline{\sigma(T) - \sigma(T|Y)}.$$

Moreover, if $Y \in SM(T)$ and $\overline{\sigma(T) - \sigma(T|Y)}$ is compact, then $\hat{T}$ is bounded.

PROOF. First, we show that, for $Y \in \text{Inv}\, T$ with $\sigma(T|Y) \neq \mathbb{C}$, $\hat{T}$ is closed. Let $\{G_0,G_1\} \in \text{cov}\, \sigma(T)$ with $G_0 \in V_\infty$ and $G_1 \in G^K$ be such that

$$(5.21) \qquad (a) \quad \sigma(T|Y) \subset G_0 \subset \overline{G}_0 \neq \mathbb{C}; \qquad (b) \quad \overline{G}_1 \subset \rho(T|Y).$$

By the 1-SDP, $X = X(T,\overline{G}_0) + \Xi(T,\overline{G}_1)$. In view of (5.21,b) and 5.12, the direct sum $\Xi(T,\overline{G}_1) \oplus X[T,\sigma(T|Y)]$ is closed. Since $Y \subset X[T,\sigma(T|Y)]$, the direct sum $\Xi(T,\overline{G}_1) \oplus Y$ is also closed.

Then

$$(5.22) \qquad X = X(T,\overline{G}_0) + [\Xi(T,\overline{G}_1) \oplus Y].$$

We infer from (5.21,a) that $Y \subset X[T,\sigma(T|Y)] \subset X(T,\overline{G}_0)$ and hence (5.22) implies the decomposition

$$X/Y = X(T,\overline{G}_0)/Y + [\Xi(T,\overline{G}_1) \oplus Y]/Y$$

where, clearly, the quotient spaces $X(T,\overline{G}_0)/Y$, $[\Xi(T,\overline{G}_1) \oplus Y]/Y$ are closed. Since

$$\sigma[T|X(T,\overline{G}_0)] \cup \sigma(T|Y) \subset \overline{G}_0 \neq \mathbb{C},$$

$\hat{T}|[X(T,\overline{G}_0)/Y]$ is closed by 3.1. Noting that $\hat{T}|[\Xi(T,\overline{G}_1) \oplus Y]/Y$ is bounded, 5.15 implies that $\hat{T}$ is closed.

Next, assume that $Y \in SM(T)$. If $\sigma(T|Y) = \mathbb{C}$ then $Y = X$ and hence $\hat{T}$ $(= \hat{0})$ is closed. If $\sigma(T|Y) \neq \mathbb{C}$ or $Y \in SM_b(T)$, (i.e. $\sigma(T|Y)$ is compact), then $\hat{T}$ is closed, by the previous part of the proof. To obtain (5.20) for $Y \in SM(T)$ or $Y \in SM_b(T)$, first suppose that $\overline{\sigma(T) - \sigma(T|Y)} \neq \mathbb{C}$. Let $G_0 \in V_\infty$ with $\overline{G}_0 \neq \mathbb{C}$ be an open neighborhood of $\overline{\sigma(T) - \sigma(T|Y)}$ and let G_1 be relatively compact such that $\{G_0,G_1\} \in \text{cov}\, \sigma(T)$ and $\overline{G}_1 \cap \overline{\sigma(T) - \sigma(T|Y)} = \emptyset$. Then

$$(5.23) \qquad \sigma[T|\Xi(T,\overline{G}_1)] \subset \overline{G}_1 \cap \sigma(T) =$$

$$\overline{G}_1 \cap \{[\sigma(T) - \sigma(T|Y)] \cup \sigma(T|Y)\} \subset \sigma(T|Y).$$

By the hypothesis on Y, $\Xi(T,\overline{G}_1) \subset Y$ and hence $\Xi(T,\overline{G}_1) \subset Y \cap \mathcal{D}_T$. By 3.2, $\hat{T}$ and $\tilde{T} = [T|X(T,\overline{G}_0)]/X(T,\overline{G}_0) \cap Y$ are similar and

$$\sigma(\hat{T}) = \sigma(\tilde{T}) \subset \sigma[T|X(T,\overline{G}_0)] \cup \sigma[T|X(T,\overline{G}_0) \cap Y] \subset \overline{G}_0.$$

Since $G_0 \in V_\infty$ is an arbitrary open neighborhood of $\overline{\sigma(T) - \sigma(T|Y)}$, inclusion

$$\sigma(\hat{T}) \subset \overline{\sigma(T) - \sigma(T|Y)}$$

follows. Since the opposite inclusion is evident, (5.20) is obtained.

Now, suppose that $\overline{\sigma(T) - \sigma(T|Y)} = \mathbb{C}$. Then, it follows from $\sigma(T) \subset \sigma(\hat{T}) \cup \sigma(T|Y)$ that $\sigma(\hat{T}) \supset \overline{\sigma(T) - \sigma(T|Y)} = \mathbb{C}$ and hence (5.20) holds in this case too.

Finally assume that $\overline{\sigma(T) - \sigma(T|Y)}$ is compact and $Y \in SM(T)$. Let $\{G_0,G_1\} \in \text{cov}\, \sigma(T)$ with $G_0 \in V_\infty$, $G_1 \in G^K$ be such that

$$\overline{G}_0 \cap \overline{\sigma(T) - \sigma(T|Y)} = \emptyset \quad \text{and} \quad \overline{\sigma(T) - \sigma(T|Y)} \subset G_1.$$

Then $X = X(T,\overline{G}_0) + \Xi(T,\overline{G}_1)$ and

(5.24) $$\sigma[T|X(T,\overline{G}_0)] \subset \sigma(T|Y),$$

analogous to (5.23), can be established. Since $Y \in SM(T)$, (5.24) implies that $X(T,\overline{G}_0) \subset Y$. Thus, it follows that

$$X = Y + \Xi(T,\overline{G}_1).$$

$\hat{T}$ and $\tilde{T} = [T|\Xi(T,\overline{G}_1)]/\Xi(T,\overline{G}_1) \cap Y$ being similar (Proposition 3.4), with the latter bounded, we conclude that $\hat{T}$ is bounded. □

Some characterizations of operators with the 1-SDP now follow.

5.17. THEOREM. Given T, the following assertions are equivalent:

(I) T has the 1-SDP;

(II) (a) T has property (κ);
(b) for every compact F, $T' = T/\Xi(T,F)$ is closed and

$$\sigma(T') \subset (\text{Int } F)^c;$$

(III) for every $G \in G^K$, there is $Y \in \text{Inv } T$ such that

(5.25) $$Y \subset \mathcal{D}_T, \quad \sigma(T|Y) \subset \overline{G}, \quad \hat{T} = T/Y \text{ is closed}, \quad \sigma(\hat{T}) \subset G^c;$$

(IV) (a) T has property (κ);
(b) for each closed $F \in V_\infty$, $T'' = T/X(T,F)$ is bounded and

$$\sigma(T'') \subset (\text{Int } F)^c;$$

(V) for every open $G \in V_\infty$, there is $Y \in \text{Inv } T$ with

$$\sigma(T|Y) \subset \overline{G}, \quad \hat{T} = T/Y \text{ bounded and } \sigma(\hat{T}) \subset G^c;$$

(VI) for every $G \in G^K$, there is $Y \in \text{Inv } T$ such that

(5.26) $$\sigma(T|Y) \subset G^c, \quad \hat{T} = T/Y \text{ is bounded and } \sigma(\hat{T}) \subset \overline{G}.$$

PROOF. The proof will be carried out through the implications:

$$(I) \Rightarrow (II) \Rightarrow (III) \Rightarrow (I) \Rightarrow (IV) \Rightarrow (V) \Rightarrow (VI) \Rightarrow (I).$$

(I) => (II): (a) follows from 5.9 and 5.6. Let F be compact. If $\text{Int}\, F = \emptyset$ then $\sigma(T') \subset \mathbb{C} = (\text{Int}\, F)^c$. For $\text{Int}\, F \neq \emptyset$, 5.14 implies

$$\overline{\text{Int}\, F \cap \sigma(T)} \subset \sigma[T|X(T,\overline{\text{Int}\, F})] \subset \sigma[T|X(T,F)] = \sigma[T|\Xi(T,F)].$$

With the help of 5.16, one obtains successively

$$\sigma(T') = \overline{\sigma(T) - \sigma[T|\Xi(T,F)]} \subset \overline{\sigma(T) - [\text{Int}\, F \cap \sigma(T)]} \subset (\text{Int}\, F)^c.$$

(II) => (III) is evident.

(III) => (I): First, we show that T has the SVEP. Let $f : \omega_f \to \mathcal{D}_T$ be analytic and verify equation

(5.27) $$(\lambda - T)f(\lambda) = 0 \text{ on an open } \omega_f \subset \mathbb{C}.$$

We assume that ω_f is connected. Select $G \in G^K$ such that $\overline{G} \subset \omega_f$. Let $Y \in \text{Inv}\, T$ satisfy (5.25). In view of (5.27),

(5.28) $$(\lambda - \hat{T})\hat{f}(\lambda) = \hat{0} \text{ on } \omega_f.$$

Since $G \subset \rho(\hat{T})$, (5.28) implies that $\hat{f}(\lambda) = \hat{0}$, i.e. $f(\lambda) \in Y$ on G. Thus $f(\lambda) \in Y$ on ω_f, by analytic continuation. It follows from (5.27) and (5.25) that $f(\lambda) = 0$ on $\omega_f - \overline{G}$ and hence $f(\lambda) = 0$ on ω_f, by analytic continuation. Consequently, T has the SVEP.

Next, let $\{G_0, G_1\} \in \text{cov}\, \sigma(T)$ with $G_0 \in V_\infty$, $G_1 \in G^K$. For $G = G_0 \cap G_1$, there is $Y \in \text{Inv}\, T$ satisfying (5.25). Then, 4.16 (i) implies that $Y \in AI(T)$ and hence $\sigma(T|Y) \subset \sigma(T)$. Then $\sigma(\hat{T}) \subset \sigma(T) \cup \sigma(T|Y) = \sigma(T)$ and by the last of (5.25), one obtains

$$\sigma(\hat{T}) \subset G^c \cap \sigma(T) = [G_0^c \cap \sigma(T)] \cup [G_1^c \cap \sigma(T)].$$

The sets $G_0^c \cap \sigma(T)$, $G_1^c \cap \sigma(T)$ are disjoint and the former is compact. By the functional calculus, there are $\hat{Z}_0, \hat{Z}_1 \in \text{Inv}\, \hat{T}$ satisfying:

$$X/Y = \hat{Z}_0 \oplus \hat{Z}_1, \quad \hat{Z}_1 \subset \mathcal{D}_{\hat{T}}, \quad \sigma(\hat{T}|\hat{Z}_i) \subset G_j^c \cap \sigma(T), \; j \neq i;\; i,j = 0,1.$$

The subspaces

$$Z_i = \{x \in X,\; x \in \hat{x},\; \hat{x} \in \hat{Z}_i\}, \quad i = 0,1$$

are invariant under T and one obtains

(5.29) $$X = Z_0 + Z_1;$$

(5.30) $$\sigma(T|Z_i) \subset \sigma(T|Y) \cup \sigma(\hat{T}|\hat{Z}_i) \subset \overline{G} \cup [G_j^c \cap \sigma(T)] \subset \overline{G}_i, \; j \neq i,\; i,j = 0,1.$$

Since $Y \subset \mathcal{D}_T$ and $\hat{Z}_1 \subset \mathcal{D}_{\hat{T}}$, it follows from the definition of Z_1 that

(5.31) $$Z_1 \subset \mathcal{D}_T.$$

Thus, T has the 1-SDP, by (5.29),(5.30) and (5.31).

(I) => (IV): T has property (κ) by 5.9 and 5.6. Since $F \in V_\infty$ and by 5.14,

(5.32) $$\overline{\text{Int } F \cap \sigma(T)} \subset \sigma[T|X(T,\overline{\text{Int } F})] \subset \sigma[T|X(T,F)],$$

it follows that $\overline{\sigma(T) - \sigma[T|X(T,F)]}$ is compact. Therefore, T" is bounded, by 5.16. Furthermore, it follows from (5.32), that

$$\sigma(T'') = \overline{\sigma(T) - \sigma[T|X(T,F)]} \subset \overline{\sigma(T) - [\text{Int } F \cap \sigma(T)]} \subset (\text{Int } F)^c.$$

(IV) => (V): For open $G \in V_\infty$, $Y = X(T,\overline{G})$ satisfies all conditions of (V).

(V) => (VI): Let $G \in G^K$ be given. Then $H = (\overline{G})^c$ is an open neighborhood of ∞ and, by hypothesis, there is $Y \in \text{Inv } T$ such that

$$\sigma(T|Y) \subset \overline{H} \subset G^c,\quad \hat{T} = T/Y \text{ is bounded and } \sigma(\hat{T}) \subset H^c = \overline{G}.$$

(VI) => (I): T has property (β) and hence property (κ), by 5.8 and 5.6. Let $\{G_0,G_1\} \in \text{cov } \mathbb{C}$ with $G_0 \in V_\infty$, $\overline{G}_0 \neq \mathbb{C}$, and $G_1 \in G^K$. Select an open $H \in V_\infty$ such that $\overline{H} \subset G_0$ and $\{H,G_1\} \in \text{cov } \mathbb{C}$. The open $G = (\overline{G}_0)^c \cup (H \cap G_1)$ is relatively compact and, by hypothesis, there exists $Y \in \text{Inv } T$ satisfying conditions (5.26). Since $\overline{\mathbb{C} - \overline{G}_0} \cap \overline{H \cap G_1} = \emptyset$, there are $\hat{Z}_0, \hat{Z}_1 \in \text{Inv } \hat{T}$ producing the decomposition

$$X/Y = \hat{Z}_0 \oplus \hat{Z}_1,\quad \sigma(\hat{T}|\hat{Z}_0) \subset \overline{H \cap G_1},\quad \sigma(\hat{T}|\hat{Z}_1) \subset \overline{\mathbb{C} - \overline{G}_0} \subset G_0^c.$$

In terms of $Z_i = \{x \in X : x \in \hat{x},\ \hat{x} \in \hat{Z}_i\} \in \text{Inv } T$ $(i=0,1)$, we have

(5.33) $$X = Z_0 + Z_1;$$

(5.34) $$\sigma(T|Z_0) \subset \sigma(\hat{T}|\hat{Z}_0) \cup \sigma(T|Y) \subset \overline{G}_0;$$

(5.35) $$\sigma(T|Z_1) \subset \sigma(\hat{T}|\hat{Z}_1) \cup \sigma(T|Y) \subset H^c \cup G_1^c.$$

It follows from (5.34) and (5.35) that $Z_0 \subset X(T,\overline{G}_0)$ and

$$Z_1 \subset X(T,H^c \cup G_1^c) = \Xi(T,H^c) \oplus X(T,G_1^c) \subset \Xi(T,\overline{G}_1) + X(T,\overline{G}_0).$$

Thus, we infer from (5.33) that

$$X = X(T,\overline{G}_0) + \Xi(T,\overline{G}_1)$$

and hence T has the SDP. □

5.18. COROLLARY. Given T with the 1-SDP, for every compact F, $\Xi(T,F)$ is analytically invariant under T.

PROOF follows directly from 5.17 (I,II), 4.19 (ii) and 4.24. □

5.19. PROPOSITION. Given T with the 1-SDP, for every open $G \in V_\infty$, $X(T,\overline{G})$ is analytically invariant under T.

PROOF. Let $f : \omega_f \to \mathcal{D}_T$ be analytic and satisfy condition

$$(\lambda-T)f(\lambda) \in X(T,\overline{G}) \quad \text{on an open } \omega_f \subset \mathbb{C}.$$

We assume that ω_f is connected. If $\omega_f \cap G \neq \emptyset$ then, with the help of 4.24, it follows easily that $f(\lambda) \in X(T,\overline{G})$ on ω_f. Hence, assume that $\omega_f \subset G^c$. Select $G_1 \in G^K$ such that $\{G,G_1\} \in \operatorname{cov} \sigma(T)$. In view of 3.11 and 5.10, for every $\lambda_0 \in \omega_f$, there is a neighborhood ω of λ_0 and there are analytic functions

$$f_0 : \omega \to X(T,\overline{G}), \quad f_1 : \omega \to \Xi(T,\overline{G}_1) \text{ with } f(\lambda) = f_0(\lambda) + f_1(\lambda)$$

on ω. Since $f(\lambda) \in \mathcal{D}_T$ and $f_1(\lambda) \in \mathcal{D}_T$, one has $f_0(\lambda) \in \mathcal{D}_T$ on ω. Moreover, $T|\Xi(T,\overline{G}_1)$ being bounded, the function $g : \omega \to X(T,\overline{G}) \cap \Xi(T,\overline{G}_1)$, defined by

$$g(\lambda) = (\lambda-T)f_1(\lambda) = (\lambda-T)[f(\lambda) - f_0(\lambda)]$$

is analytic on ω. It follows from the SVEP and

$$\sigma[T|X(T,\overline{G}) \cap \Xi(T,\overline{G}_1)] = \sigma[T|\Xi(T,\overline{G} \cap \overline{G}_1)] \subset \overline{G}$$

that

$$f_1(\lambda) = R[\lambda;T|\Xi(T,\overline{G} \cap \overline{G}_1)]g(\lambda) \in \Xi(T,\overline{G} \cap \overline{G}_1) \subset X(T,\overline{G}) \quad \text{on } \omega.$$

Therefore, $f(\lambda) \in X(T,\overline{G})$ on ω, and hence $f(\lambda) \in X(T,\overline{G})$ on ω_f, by analytic continuation. □

5.20. COROLLARY. Given T with the 1-SDP, for every closed F, $X(T,F)$ is analytically invariant under T.

PROOF is left to the reader. □

The concept of *decomposable operator* brings the power of the spectral maximal spaces to bear in the spectral decomposition problem. Basically, a decomposable operator produces a decomposition of the underlying Banach space into a linear sum of spectral maximal spaces. In the light of the independent works [A.1979], [N.1978], [L.1981] and [Sh.1981], the decomposable operators and the operators with the SDP are indistinguishable within the Banach algebra B(X). For unbounded operators, however, the two classes of operators no longer coincide.

The remainder of this section tackles the relationship of these two classes of operators, in the framework of the two-summand decomposition of the underlying space. The more meaningful results on this problem are deferred until we reach the conclusion of the next section, when the two-summand decomposition will be unified with the general spectral decomposition problem.

5.21. DEFINITION. Given $n \in \mathbb{N}$, T is said to be *n-decomposable* if, for every $\{G_i\}_{i=0}^n \in \operatorname{cov} \sigma(T)$ with $G_0 \in V_\infty$, $G_i \in \mathcal{G}^K$, there is a system $\{X_i\}_{i=0}^n \subset SM(T)$ satisfying conditions (I), (II) and (III) of 5.1. If, for every $n \in \mathbb{N}$, T is n-decomposable, then T is said to be *decomposable*.

It is known that, contrary to the bounded case, for a linear operator T with the SVEP, $\sigma(x,T) = \emptyset$ does not necessarily imply that $x = 0$, as illustrated by the following

5.22. EXAMPLE. Let $Y = Z = C[0,1]$ be the Banach space of all continuous $\mathbb{C}$-valued functions defined on the unit interval. Let A be a decomposable multiplication operator on Y, i.e.

$$(Ay)(t) = ty(t), \quad y \in Y, \quad t \in [0,1],$$

and define the operator $B : \mathcal{D}_B(\subset Z) \to Z$, by

$$(Bz)(t) = \frac{dz}{dt}(t), \quad \mathcal{D}_B = \{z \in Z : \frac{dz}{dt}(t) \in Z,\ z(0) = 0\}.$$

Then B is a closed operator with $\sigma(B) = \emptyset$ (see [Du-S.1967; VII.10. Example 1(a)]). Let

$$X = Y \oplus Z \quad \text{and} \quad T = A \oplus B.$$

Then $\sigma(T) = \sigma(A) = [0,1]$. Let $\{G_i\}_{i=0}^n \in \operatorname{cov} \sigma(T)$ with $G_0 \in V_\infty$, $G_i \in \mathcal{G}^K$. Since A is decomposable, we have

$$Y = \sum_{i=0}^n Y(A,\overline{G}_i), \quad Y(A,\overline{G}_i) \subset \mathcal{D}_T \quad (1 \le i \le n);$$

$$\sigma[A|Y(A,\overline{G}_i)] \subset \overline{G}_i, \quad 0 \le i \le n.$$

Put

$$X_0 = Y(A,\overline{G}_0) \oplus Z, \quad X_i = Y(A,\overline{G}_i) \quad \text{for } 1 \le i \le n$$

and obtain

$$X = Y \oplus Z = \sum_{i=0}^{n} X_i, \quad \sigma(T|X_i) \subset \overline{G}_i, \quad 0 \le i \le n.$$

Thus T has the SDP. On the other hand, for any $x \in Z$,

$$\sigma(x,T) \subset \sigma(T|Z) = \sigma(B) = \emptyset. \ \square$$

5.23. THEOREM. Given T, the following assertions are equivalent:

(I) T is 1-decomposable;

(II) T has the 1-SDP and $X(T,\emptyset) = \{0\}$, or

T has the 1-SDP and $\{0\} \in SM(T)$;

(III) T has the 1-SDP and every T-bounded spectral maximal space is a spectral maximal space of T;

(IV) T has the 1-SDP and $X(T,F) \subset \mathcal{D}_T$ for some compact F.

PROOF. The conclusion will be reached through the following sequel of implications: (I) => (II) => (III) => (IV) =>(II) and (III) => (I).

(I) => (II): Clearly, T has the 1-SDP. Let $\{G_0,G_1\} \in \mathrm{cov}\ \sigma(T)$ with $G_0 \in V_\infty$, $G_1 \in G^K$. The decomposition of X can be written in the following terms:

$$X = X_0 + X_1 \quad \text{with} \quad X_i \in SM(T) \quad \text{and} \quad X_1 \subset \mathcal{D}_T.$$

Then $\sigma[T|X(T,\emptyset)] = \emptyset \subset \sigma(T|X_1)$ implies that $X(T,\emptyset) \subset X_1 \subset \mathcal{D}_T$. Now 4.33 applied to the space $Y = X(T,\emptyset)$ asserts that each $x \in X(T,\emptyset)$ is the zero vector and hence $X(T,\emptyset) = \{0\}$. Equivalently, $\{0\} \in SM(T)$.

(II) => (III): follows from 4.31.

(III) => (IV): Let $F \subset \mathbb{C}$ be compact. It follows from the hypothesis that the T-bounded spectral maximal space $\Xi(T,F)$ is spectral maximal. Consequently, $\sigma[T|X(T,F)] = \sigma[T|\Xi(T,F)]$ implies that

$$X(T,F) \subset \Xi(T,F) \subset \mathcal{D}_T.$$

(IV) => (II): If, for some compact $F \subset \mathbb{C}$, $X(T,F) \subset \mathcal{D}_T$, then

$$X(T,F) = \Xi(T,F) \oplus X(T,\emptyset) \quad \text{and} \quad \Xi(T,F) \subset \mathcal{D}_T$$

imply that $X(T,\emptyset) = \{0\}$.

(III) => (I): For $\{G_0,G_1\} \in \mathrm{cov}\ \sigma(T)$ with $G_0 \in V_\infty$ and $G_1 \in G^k$, $X = X(T,\overline{G}_0) + \Xi(T,\overline{G}_1)$. Since, by hypothesis, $\Xi(T,\overline{G}_1) \in SM(T)$,

T is 1-decomposable. $\square$

5.24. COROLLARY. Let T be 1-decomposable. If, for every $x \in X$, $\sigma(x,T)$ is compact then T is bounded.

PROOF. Let $x \in X$ be arbitrary. It follows from 5.23 (II) and 4.34 that, for every compact $F \subset \mathbb{C}$, $X(T,F) = \Xi(T,F) \subset \mathcal{D}_T$. In particular, for $F = \sigma(x,T)$, we have

$$x \in X[T,\sigma(x,T)] = \Xi[T,\sigma(x,T)] \subset \mathcal{D}_T$$

and hence $\mathcal{D}_T = X$. $\square$

If the SDP (or the property to be decomposable) of a closed operator T is inducible to its restrictions, by spectral maximal spaces, then T enjoys more powerful spectral properties. One such property is the "local version" of (5.20). We shall devote §.15 to the study of such operators. For the time being, we establish a consequence of the property, we alluded above, to be used later.

5.25. PROPOSITION. Suppose that T has property (κ) and, for every T-bounded spectral maximal space Y, $\hat{T} = T/Y$ is closed and

$$\sigma(\hat{x},\hat{T}) = \overline{\sigma(x,T) - \sigma(T|Y)}, \tag{5.36}$$

where $x \in X$, $\hat{x} = x + Y \in X/Y$. Then, for any closed F, $\hat{X}(\hat{T},F)$ is closed and, for $\hat{x} \in \hat{X}(\hat{T},F)$, we have

$$\sigma(\hat{x},\hat{T}) = \sigma[\hat{x},\hat{T}|\hat{X}(\hat{T},F)], \tag{5.37}$$

where $\hat{X}(\hat{T},\cdot)$ is the spectral manifold of $\hat{T}$ in $\hat{X} = X/Y$.

PROOF. Given a closed $F \subset \mathbb{C}$, let $\hat{x} \in \hat{X}(\hat{T},F)$. For every $x \in \hat{x}$, (5.36) implies

$$\sigma(x,T) \subset \sigma(\hat{x},\hat{T}) \cup \sigma(T|Y) \subset F \cup \sigma(T|Y)$$

and hence $x \in X[T,F \cup \sigma(T|Y)]$. Conversely, for every $x \in X[T,F \cup \sigma(T|Y)]$, (5.36) implies

$$\sigma(\hat{x},\hat{T}) = \overline{\sigma(x,T) - \sigma(T|Y)} \subset \overline{F \cup \sigma(T|Y) - \sigma(T|Y)} \subset F$$

and hence $\hat{x} \in \hat{X}(\hat{T},F)$. Thus, the following equality holds

$$\hat{X}(\hat{T},F) = X[T,F \cup \sigma(T|Y)]/Y.$$

Consequently, $\hat{X}(\hat{T},F)$ is closed for closed F. Moreover, since for every $\hat{x} \in \hat{X}(\hat{T},F)$, we have $\sigma[\hat{x}(\lambda),\hat{T}] = \sigma(\hat{x},\hat{T}) \subset F$, it follows that

$$\hat{x}(\lambda) \in \hat{X}(\hat{T},F) \quad \text{for all } \lambda \in \rho(\hat{x},\hat{T}).$$

By 4.8, $\hat{X}(\hat{T},F)$ is a μ-space of $\hat{T}$ and hence (5.37) holds. $\square$

§.6. THE EQUIVALENCE OF THE 1-SDP AND THE SDP.

We are now in a position to fulfill our earlier promise to make possible the study of the general spectral decomposition problem in terms of the special 1-SDP.

6.1. LEMMA. Let T have the 1-SDP and let $x \in X$. If there is a sequence $\{f_n : G \to \mathcal{D}_T\}$ of analytic functions defined on a set $G \in G^K$ such that

$$(6.1) \qquad \|x - (\lambda - T)f_n(\lambda)\| \to 0 \quad (\text{as} \quad n \to \infty)$$

uniformly on G, then $G \subset \rho(x,T)$.

PROOF. The double sequence $\{g_{mn} = f_m - f_n\}_{m,n}$ is such that

$$(\lambda - T)g_{mn}(\lambda) \to 0$$

uniformly on G. By 5.9, T has property (β) and hence $g_{mn}(\lambda) \to 0$ uniformly on G. Consequently, the limit function $f(\lambda) = \lim_n f_n(\lambda)$ is analytic on G. T being closed, (6.1) implies

$$(\lambda - T)f(\lambda) = x \quad \text{on } G$$

and hence $G \subset \rho(x,T)$. □

An essential link between the 1-SDP and the SDP of the given T now follows.

6.2. THEOREM. Given T with the 1-SDP, let $F \subset \mathbb{C}$ be closed. For any $\{G_i\}_{i=0}^m \in \text{cov } F$ with $G_0 \in V_\infty$ and $\{G_i\}_{i=1}^m \subset G^K$, we have

$$(6.2) \qquad X(T,F) \subset X(T,\overline{G}_0) + \sum_{i=1}^m \Xi(T,\overline{G}_i).$$

PROOF. We shall establish (6.2) for m=1, in three parts. A subsequent induction on m, will bring the proof to the end.

Part A. We build up a sum representation of an arbitrary $x \in X(T,F)$. Let $\{G_0,G_1\} \in \text{cov } F$ with $G_0 \in V_\infty$, $G_1 \in G^K$. Put $K = \overline{G_0 \cap G_1}$ and deduce from 5.17 that $\hat{T} = T/\Xi(T,K)$ is closed with $\sigma(\hat{T}) \subset (G_0 \cap G_1)^c$. Let $x \in X(T,F)$. Since, for $\lambda \in F^c \subset \rho[T|X(T,F)]$, we have $x(\lambda) = R[\lambda;T|X(T,F)]x$, the canonical map $X \to X/\Xi(T,K)$ gives

$$(\lambda - \hat{T})\hat{x}(\lambda) = \hat{x} \text{ on } F^c \quad \text{and} \quad (\lambda - \hat{T})R(\lambda;\hat{T})\hat{x} = \hat{x} \text{ on } G_0 \cap G_1.$$

Consequently, the function

$$(6.3)\qquad \hat{f}(\lambda) = \begin{cases} \hat{x}(\lambda), & \text{if } \lambda \in F^c \\ R(\lambda;\hat{T})\hat{x}, & \text{if } \lambda \in G_0 \cap G_1 \end{cases}$$

is well-defined on $F^c \cup (G_0 \cap G_1)$. Consider a bounded Cauchy neighborhood Δ of $F - G_0$ with $\bar{\Delta} \cap (F - G_1) = \emptyset$. Define $\hat{x}_0$ and $\hat{x}_1$, as follows

$$(6.4)\qquad \hat{x}_1 = \frac{1}{2\pi i}\int_{\partial\Delta} \hat{f}(\lambda)d\lambda \quad\text{and}\quad \hat{x}_0 = \hat{x} - \hat{x}_1.$$

Clearly, $\hat{x}_1$ is independent of the choice of Δ. By (6.4), we have

$$(6.5)\qquad x = x_0 + x_1 + y \quad\text{with}\quad x_0 \in \hat{x}_0,\ x_1 \in \hat{x}_1,\ y \in \Xi(T,K).$$

Part B. In this part, we show that $x_i \in X(T,\bar{G}_i)$, $i=0,1$. For $i=1$, let $\lambda_0 \in (\bar{G}_1)^c$. There is a neighborhood δ of λ_0 such that $\bar{\delta} \cap \bar{G}_1 = \emptyset$. Choose a Cauchy domain Δ such that $\bar{\Delta} \cap \bar{\delta} = \emptyset$. Since $\hat{T}$ is closed, for $\lambda \in \delta$, we obtain successively

$$(6.6)\qquad (\lambda-\hat{T})\,\frac{1}{2\pi i}\int_{\partial\Delta}\frac{\hat{f}(\mu)}{\lambda-\mu}\,d\mu = \frac{1}{2\pi i}\int_{\partial\Delta}\frac{(\lambda-\hat{T})\hat{f}(\mu)}{\lambda-\mu}\,d\mu$$

$$= \frac{1}{2\pi i}\int_{\partial\Delta}\frac{\hat{x}}{\lambda-\mu}\,d\mu + \frac{1}{2\pi i}\int_{\partial\Delta}\hat{f}(\mu)d\mu = \hat{x}_1.$$

It follows from 5.18 and 4.14 that $\hat{T}$ has the SVEP and hence (6.6) implies that $\sigma(\hat{x}_1,\hat{T}) \subset \bar{G}_1$. Furthermore, by 4.35,

$$\sigma(x_1,T) \subset \{\sigma(x_1,T) \cap \sigma[T|\Xi(T,K)]\} \cup \sigma(\hat{x}_1,\hat{T})$$

$$\subset \sigma[T|\Xi(T,K)] \cup \sigma(\hat{x}_1,\hat{T}) \subset \bar{G}_1$$

and hence $x_1 \in X(T,\bar{G}_1)$. In a similar way, $x_0 \in X(T,\bar{G}_0)$ follows easily.

Part C. In this final part, we shall obtain (6.2). Note that $\Xi(T,K) \subset X(T,\bar{G}_0)$, use (6.5) and the conclusions of Part B, to write

$$X(T,F) \subset X(T,\bar{G}_0) + X(T,\bar{G}_1).$$

It follows from $X(T,\bar{G}_1) = \Xi(T,\bar{G}_1) \oplus X(T,\emptyset)$ and $X(T,\emptyset) \subset X(T,\bar{G}_0)$ that $X(T,\bar{G}_0) + X(T,\bar{G}_1) = X(T,\bar{G}_0) + \Xi(T,\bar{G}_1)$ and hence (6.2) holds, for $m=1$.

We proceed by induction on m. Let $\{G_i\}_{i=0}^{m+1} \in \text{cov } F$ with $G_0 \in V_\infty$, $\{G_i\}_{i=1}^{m+1} \subset G^K$. Let H be open with $F \subset H \cup G_{m+1}$, $\bar{H} \subset \bigcup_{i=0}^{m} G_i$. Then, we have

$$(6.7)\qquad X(T,F) \subset X(T,\bar{H}) + \Xi(T,\bar{G}_{m+1})$$

and the hypothesis of the induction implies

$$(6.8) \qquad X(T,\overline{H}) \subset X(T,\overline{G}_0) + \sum_{i=1}^{m} \Xi(T,\overline{G}_i).$$

Thus, it follows from (6.7) and (6.8), that

$$X(T,F) \subset X(T,\overline{G}_0) + \sum_{i=1}^{m+1} \Xi(T,\overline{G}_i). \ \square$$

6.3. COROLLARY. If T has the 1-SDP then T has the SDP. Moreover, for any $\{G_i\}_{i=0}^{n} \in \operatorname{cov}\sigma(T)$ with $G_0 \in V_\infty$ and $\{G_i\}_{i=1}^{n} \subset G^K$, the following decomposition holds

$$(6.9) \qquad X = X(T,\overline{G}_0) + \sum_{i=1}^{n} \Xi(T,\overline{G}_i).$$

PROOF. For $F = \sigma(T)$, $X(T,F) = X$, 6.2 implies (6.9). $\square$

For bounded operators, the equivalence of decomposable and two-decomposable (the latter corresponds to 1-decomposable operators in our terminology) was proved in [R.1978]. With the help of 5.23 and 6.3, we can extend such equivalence to closed operators.

6.4. THEOREM. Given T, the following assertions are equivalent:

(I) T is 1-decomposable;

(II) T has the SDP and $X(T,\emptyset) = \{0\}$, or

T has the SDP and $\{0\}$ is a spectral maximal space of T;

(III) T has the SDP and every T-bounded spectral maximal space is a spectral maximal space of T;

(IV) T has the SDP and $X(T,F) \subset \mathcal{D}_T$ for some compact F;

(V) T is decomposable.

PROOF. In view of 5.23 and 6.3, it suffices to show that one of the conditions (I) - (IV) is equivalent to (V). Since the implication (V) => (II) is obvious, we shall show that (III) implies (V). Since T has the SDP, for every $\{G_i\}_{i=0}^{n} \in \operatorname{cov}\sigma(T)$, with $G_0 \in V_\infty$, $\{G_i\}_{i=1}^{n} \subset G^K$, we have (6.9). By hypothesis, $\Xi(T,\overline{G}_i) \in SM(T)$, $1 \le i \le n$, and therefore, T is decomposable. $\square$

6.5. COROLLARY. Given $T \in B(X)$, the following assertions are equivalent.

(I) T has the 1-SDP;

(II) T has the SDP;

(III) T is 1-decomposable;

(IV) T is decomposable.

PROOF. We only have to obtain implication (II) => (IV). Since $X(T,\emptyset) \subset \mathcal{D}_T$, we have $X(T,\emptyset) = \{0\}$ and hence (II) => (IV) follows from 6.4. □

§.7. SPECTRAL ELEMENTS IN FUNCTIONAL CALCULUS.

In this section, we indulge in some functional calculus which stems from the theory of closed operators. Our main purpose is to show that if T has the SDP then f(T) is decomposable and, conversely, if f is nonconstant on every component of its domain which intersects the spectrum of T, then f(T) decomposable implies that T has the SDP.

Throughout this section, we *assume* that $\rho(T) \neq \emptyset$. For the reader's convenience, 2.8, 2.9 and 2.10 should be reviewed. Also the hyperinvariance of both spectral maximal and T-bounded spectral maximal spaces, as expressed by 4.23, will be referred to.

Given T, fix $\alpha \in \rho(T)$. Define the map $\Phi : \mathbb{C}_\infty \to \mathbb{C}_\infty$, by

$$\Phi(\lambda) = \begin{cases} (\lambda-\alpha)^{-1}, & \text{if } \lambda \neq \alpha, \quad \lambda \neq \infty; \\ 0, & \text{if } \lambda = \infty; \\ \infty, & \text{if } \lambda = \alpha. \end{cases}$$

Denote $A = (T-\alpha)^{-1}$.

7.1. LEMMA. If A is decomposable, then

(i) for every closed F, $X(A,F) \in \operatorname{Inv} T$;

(ii) if $0 \notin F$, then $X(A,F) \subset \mathcal{D}_T$.

PROOF. We run the proof in three parts.

Part I. In this part, we show that, for closed F with $0 \notin F$,

$$(7.1) \qquad X(A,F) \in \operatorname{Inv} T \quad \text{and} \quad X(A,F) \subset \mathcal{D}_T.$$

It follows from condition $0 \notin F$ that $[A|X(A,F)]^{-1}$ is bounded. Let $x \in X(A,F)$. Then $x = A[A|X(A,F)]^{-1}x \in \mathcal{R}_A = \mathcal{D}_T$. Since A is injective, it follows from

$$A\{(T-\alpha)x - [A|X(A,F)]^{-1}x\} = x - x = 0,$$

that

$$(7.2) \qquad (T-\alpha)x = [A|X(A,F)]^{-1}x \in X(A,F)$$

and hence $X(A,F) \in \text{Inv } T$. Thus, (7.1) is proved.

Part II. In this part we show that, for every open G which contains 0, $X(A,\overline{G}) \in \text{Inv } T$. Let $G \in \mathcal{G}$, contain 0. Choose $G_1 \in G^K$ such that $\{G,G_1\} \in \text{cov } \sigma(A)$ and $0 \notin \overline{G}_1$. Since A is decomposable,

$$(7.3) \qquad X = X(A,\overline{G}) + X(A,\overline{G}_1).$$

Let $x \in X(A,\overline{G}) \cap \mathcal{D}_T$. In view of (7.3), there is a representation

$$(T-\alpha)x = y_0 + y_1 \quad \text{with} \quad y_0 \in X(A,\overline{G}),\ y_1 \in X(A,\overline{G}_1).$$

Let $z_1 = Ay_1$. Then $z_1 \in X(A,\overline{G}_1)$, furthermore,

$$z_1 = A[(T-\alpha)x - y_0] = x - Ay_0 \in X(A,\overline{G})$$

and hence $z_1 \in X(A,\overline{G}) \cap X(A,\overline{G}_1) = X(A,\overline{G} \cap \overline{G}_1)$. Since $0 \notin \overline{G} \cap \overline{G}_1$, it follows from Part I, that $X(A,\overline{G} \cap \overline{G}_1) \in \text{Inv } T$. Consequently,

$$y_1 = (T-\alpha)z_1 \in X(A,\overline{G} \cap \overline{G}_1) \subset X(A,\overline{G})$$

implies that

$$(T-\alpha)x = y_0 + y_1 \in X(A,\overline{G})$$

and hence $X(A,\overline{G}) \in \text{Inv } T$.

Part III. In this part, we complement the result of Part I, by showing that, for closed F containg 0, $X(A,F) \in \text{Inv } T$. Indeed, let F be closed such that $0 \in F$. Choose open sets $\{G_n\}_{n=1}^{\infty}$ such that $F \subset G_n$ for all n, and $F = \bigcap_{n=1}^{\infty} \overline{G}_n$. It follows from Part II and from

$$X(A,F) = \bigcap_{n=1}^{\infty} X(A,\overline{G}_n),$$

that $X(A,F) \in \text{Inv } T$. □

7.2. THEOREM. T has the SDP iff A is decomposable.

PROOF. (Only if): Assume that T has the SDP. Let $\{G_i\}_{i=0}^{n} \in \text{cov } \sigma(A)$ and, without loss of generality, assume that $0 \in G_0$ and $0 \notin \overline{G}_i$ $(1 \le i \le n)$. Put $H_i = \Phi^{-1}(G_i)$, $(0 \le i \le n)$, note that $H_0 \in V_\infty$ and $\{H_i\}_{i=1}^{n} \subset G^K$. By the spectral mapping theorem, $\sigma(A) = \Phi[\sigma_\infty(T)]$. Consequently,

$\{H_i\}_{i=0}^n \in \operatorname{cov} \sigma(T)$. By the SDP of T,

$$X = X(T,\overline{H}_0) + \sum_{i=1}^n \Xi(T,\overline{H}_i). \tag{7.4}$$

Since the subspaces $X(T,\overline{H}_0)$, $\Xi(T,\overline{H}_i)$, $(1\le i\le n)$ are hyperinvariant under T, they are invariant under A. Quoting again the spectral mapping theorem, we obtain

$$\sigma[A|X(T,\overline{H}_0)] = \Phi(\sigma[T|X(T,\overline{H}_0)] \cup \{\infty\}) \subset \Phi(\overline{H}_0) = \overline{G}_0; \tag{7.5}$$

$$\sigma[A|\Xi(T,\overline{H}_i)] = \Phi(\sigma[T|\Xi(T,\overline{H}_i]) \subset \Phi(\overline{H}_i) = \overline{G}_i, \quad 1\le i\le n. \tag{7.6}$$

By (7.4), (7.5), (7.6), A has the SDP and since it is bounded, A is decomposable.

(If): Assume that A is decomposable. Let $\{H_i\}_{i=0}^n \in \operatorname{cov} \sigma(T)$ with $H_0 \in V_\infty$ and $\{H_i\}_{i=1}^n \subset G^K$. Put $G_i = \Phi(H_i)$ for $0\le i\le n$, and note that $\{G_i\}_{i=0}^n \in \operatorname{cov} \sigma(A)$. Then, the hypothesis on A produces the decomposition

$$X = \sum_{i=0}^n X(A,\overline{G}_i). \tag{7.7}$$

It follows from 7.1, that $X(A,\overline{G}_i) \in \operatorname{Inv} T$ for $0\le i\le n$, and $T|X(A,\overline{G}_i)$ is bounded for $1\le i\le n$. Moreover,

$$\sigma[T|X(A,\overline{G}_i)] = \Phi^{-1}(\sigma[A|X(A,\overline{G}_i)]) \subset \Phi^{-1}(\overline{G}_i) = \overline{H}_i, \quad 0\le i\le n. \tag{7.8}$$

Thus, T has the SDP, by (7.7) and (7.8). □

7.3. COROLLARY. If T has the SDP then, for every $f \in A_T$, $f(T)$ is decomposable. Conversely, let $f \in A_T$ be nonconstant on every component of its domain which intersects $\sigma(T)$. If $f(T)$ is decomposable then T has the SDP.

PROOF. First, assume that T has the SDP. Given $f \in A_T$, let $g(\mu) = f[\Phi^{-1}(\mu)]$. By the functional calculus, $g(A) = f(T)$. Then A is decomposable, by 7.2. In this case, $g(A)$ is decomposable, as it follows from the theory of bounded decomposable operators [C-Fo.1968; Theorem 2.11]. Consequently, $f(T)$ is decomposable.

Next, assume that $f \in A_T$ is nonconstant on every component of its domain which intersects $\sigma(T)$, and let $f(T)$ be decomposable. Then, by a result of Apostol [Ap.1968,b; Theorem 3.3], A is decomposable. Thus, T has the SDP, by 7.2. □

7.4. COROLLARY. Given T with the SDP, for every $f \in A_T$ and closed F, we have

$$X[f(T),F] = \hat{X}[T,f^{-1}(F) \cap \sigma(T)], \tag{7.9}$$

where

$$\hat{X}[T,f^{-1}(F) \cap \sigma(T)] = \begin{cases} X[T,f^{-1}(F) \cap \sigma(T)], & \text{if } f(\infty) \in F; \\ \Xi[T,f^{-1}(F) \cap \sigma(T)], & \text{if } f(\infty) \notin F. \end{cases}$$

PROOF. Let F be closed. If $f(\infty) \in F$, then it follows from 2.10 that

$$X[f(T),F] = X[T,f^{-1}(F)] = X[T,f^{-1}(F) \cap \sigma(T)].$$

Next, assume that $f(\infty) \notin F$. Then $f^{-1}(F)$ is compact and 2.10 implies

$$X[f(T),F] \subset X[T,f^{-1}(F) \cap \sigma(T)]. \tag{7.10}$$

Without loss of generality, we may assume that $f(\infty) = 0$. Then $R_{f(T)} \subset \mathcal{D}_T$, where $R_{f(T)}$ denotes the range of $f(T)$. Furthermore, since $f(\infty) = 0 \in F^c$, $f(T)|X[f(T),F]$ is invertible and

$$X[f(T),F] = f(T)\{X[f(T),F]\} \subset R_{f(T)}.$$

Hence, it follows that

$$X[f(T),F] \subset \mathcal{D}_T. \tag{7.11}$$

Moreover, since $f(T)$ is decomposable, by 7.3, $X[f(T),F]$ is a spectral maximal space of $f(T)$ and hence it is invariant under T. It follows from (7.10) that, for each $x \in X[f(T),F]$,

$$\sigma(x,T) \subset f^{-1}(F). \tag{7.12}$$

$T|X[f(T),F]$ being bounded by (7.11), we have

$$\lim_{\lambda\to\infty} x_T(\lambda) = 0. \tag{7.13}$$

In view of (7.12) and (7.13), 5.11 implies that $x \in \Xi[T,f^{-1}(F)]$ and hence

$$X[f(T),F] \subset \Xi[T,f^{-1}(F)] = \Xi[T,f^{-1}(F) \cap \sigma(T)]. \tag{7.14}$$

To obtain the opposite inclusion, note that $\Xi[T,f^{-1}(F) \cap \sigma(T)]$ being a T-bounded spectral maximal space, is hyperinvariant under T and hence invariant under $f(T)$. Denote $S = T|\Xi[T,f^{-1}(F) \cap \sigma(T)]$. Since

$$\sigma(S) \subset f^{-1}(F) \cap \sigma(T) \subset \sigma(T),$$

it follows that $f \in A_S$ and

$$\sigma[f(S)] = f[\sigma(S)] \subset f[f^{-1}(F) \cap \sigma(T)] \subset F. \tag{7.15}$$

We have

$$f(S) = f\{T|\Xi[T,f^{-1}(F) \cap \sigma(T)]\} = f(T)|\Xi[T,f^{-1}(F) \cap \sigma(T)]$$

and hence (7.15) implies

$$\sigma\{f(T)|\Xi[T,f^{-1}(F) \cap \sigma(T)]\} \subset F.$$

Consequently, for every $x \in \Xi[T,f^{-1}(F) \cap \sigma(T)]$, we have $\sigma[x,f(T)] \subset F$ and this implies that

$$\Xi[T,f^{-1}(F) \cap \sigma(T)] \subset X[f(T),F]. \tag{7.16}$$

(7.16) coupled with (7.14), concludes the proof. □

7.5. COROLLARY. Given T with the SDP, if $f \in A_T$ is injective on its domain ω_f, then

$$X(T,\emptyset) = X[f(T),\{f(\infty)\}].$$

NOTES AND COMMENTS

The spectral decomposition property originates from [E-L.1978] and its unbounded variant appeared in [E.1980,a] and subsequently in [E-W.1984,a]. When introduced in 1978, the theory of the decomposable operators and that of the generalized spectral operators were at the center stage of the contemporary spectral theory. The purpose behind the introduction of the SDP was outlined in the introductory remarks of the original paper: "In the process of extending the spectral theory to ever more general linear operators, it was found that all types of operators known to perform a spectral decomposition of the underlying space, possess the single valued extension property. It is then natural to ask whether this property is or is not an intrinsic element of the spectral decomposition. In order to answer this question, we defined axiomatically the spectral decomposition of continuous linear operators on a general Banach space. Then we obtained an affirmative answer. A minimal requirement imposed on operators by any spectral theory is the existence of proper invariant subspaces. Assuming this we did not require from the invariant subspaces any specific property..." The interest in the SDP was enhanced by the discovery by Albrecht, Nagy, Lange and Shulberg, as mentioned earlier, that the SDP of a bounded operator redefines, in equivalent terms, the concept of decomposable operator. The differentiation of the two concepts starts in the case of unbounded operators. In his excellent book [V.1982], Vasilescu employed the SDP, under the name of *decomposition property* [ibid. IV. Definition 4.12], but without mentioning its source.

We show that 5.2 (a) is compatible with 5.1. Obviously, any cover of $\mathbb{C}$ is a cover of $\sigma(T)$. Conversely, let $\{G_0,G_1\} \in \operatorname{cov} \sigma(T)$ with $G_0 \in V_\infty$ and $G_1 \in G^K$. Choose $\{G_0',G_1'\} \in \operatorname{cov} \mathbb{C}$ such that $G_0' \in V_\infty$ with $\overline{G_0'} \subset G_0 \cup \rho(T)$ and $\overline{G_1'} \subset G_1$. In view of 5.10, we have

$$X = X(T,\overline{G_0'}) + \Xi(T,\overline{G_1'}).$$

Furthermore, the above mentioned equivalence statement is concluded by

$$\sigma[T|X(T,\overline{G_0'})] \subset \overline{G_0'} \cap \sigma(T) \subset [G_0 \cup \rho(T)] \cap \sigma(T) \subset G_0;$$

$$\sigma[T|\Xi(T,\overline{G_1'})] \subset \overline{G_1'} \subset G_1.$$

Property (κ), as mentioned in the text, appeared among Dunford's conditions for an operator to be spectral. Radjabalipour referred to it as *the closure condition* of the operator T, [R.1978,a]. Property (β), introduced by Bishop [Bi.1959], was successfully used by Foiaş [Fo.1968] in providing a characterization of decomposable operators in terms of *spectral capacities*. The same property played a central role in extending a characterization of the spectral maximal spaces pertaining to $\mathcal{A}$-scalar operators (from the theory of the generalized spectral operators) to spectral maximal spaces of decomposable operators [Fo.1970]. Property (β) was the main topic of Snader's doctoral thesis [Sn], [Sn.1984].

A reference for 5.6 is [V.1982, IV. Lemma 4.16]. In contrast to [ibid. IV. 4.16-4.18], in 5.5, 5.6 and further applications of property (β), we do not require that Tf_n be analytic for all n. The properties expressed by 5.7 - 5.9 were selected from [W-E] and 5.10 - 5.13 appeared in [E-W.1984,a], 5.12 being a generalization of [Ap.1968,b; Lemma 2.3]. Theorem 5.16, proved in [W-Sun], is a generalization of [Ap.1968,c; Proposition 3.3]. Partial conclusions of 5.17, confined to bounded decomposable operators were obtained in [R.1978,a]; 5.17 in the actual form is given in [W-E]. Condition (II) of 5.23 appeared in [Wa.1981].

The n-decomposable operator concept originates from [Pl.1970]. The interplay between unbounded operators with the 1-SDP (extended to the SDP, in §.6) and the corresponding 1-decomposable as well as decomposable operators (5.23, 6.4, 6.5) appeared in [W-E.a].

The basic result of §.6, i.e. 6.2 and 6.3, as proved in this text, appeared in [W.1984]. Another proof was given by Nagy in [N.1981].

Nagy used 4.15 instead of 4.35, as outlined in the following. With $\hat{f}$ defined by (6.3) and $\hat{x}_1$ expressed by (6.4), it follows from 4.15 that there exists an open $\delta' \subset \delta$ and a function $h : \delta' \to \mathcal{D}_T$ such that both h and $(\lambda-T)h(\lambda)$ are analytic on δ' and there corresponds on $X/\Xi(T,K)$:

$$\hat{h}(\lambda) = \frac{1}{2\pi i} \int_{\partial\Delta} \frac{\hat{f}(\mu)}{\lambda-\mu} d\mu \quad \text{for } \lambda \in \delta'.$$

In view of (6.6), one has

$$(\lambda-\hat{T})\hat{h}(\lambda) = \hat{x}_1, \ \lambda \in \delta'$$

or, equivalently,

$$(\lambda-T)h(\lambda) = x_1 + g(\lambda),$$

where $g : \delta' \to \Xi(T,K)$ is analytic on δ'. Since $\overline{\delta'} \cap \overline{G}_1 = \emptyset$, one finds

$$(\lambda-T)\{h(\lambda) - R[\lambda;T|\Xi(T,K)]g(\lambda)\} = x_1$$

and hence $\lambda_0 \in \rho(x_1,T)$. By the choice of λ_0, one obtains $\sigma(x_1,T) \subset \overline{G}_1$ and therefore, $x_1 \in X(T,\overline{G}_1)$. This, and a similar proof for $x_0 \in X(T,\overline{G}_0)$, concludes Part B of the proof.

It is well known that the spectral decomposition problem and the functional calculus are closely related. An operator T, which produces a spectral decomposition of the underlying space X, admits a functional calculus which preserves that property of T. Conversely, however, an operator may possess quite a satisfactory functional calculus, without the ability of decomposing X. If f(T) decomposes the space X, then extra conditions are needed for T to belong to the same class of operators. For normal and spectral operators, this problem has been under the scrutiny of a variety of works, such as [Ba.1954], [Pu.1957], [St.1962], [Ku.1962], [Ku.1963], [K.1965], [Du.1966]. For bounded decomposable operators, we mention the papers [C-Fo.1967], [Ap.1968], [Ap.1968,b], [Ap.1968,c]. For finite systems of commuting linear operators, some functional calculus related works are [Tay.1970], [A.1974], [V.1978] and [Es.1981]. Summing up the relationship between the spectral decomposition problem and the functional calculus for bounded operators: if T is a bounded operator with the SDP (i.e. decomposable), then f(T) has the SDP [E-L.1978]. For the converse property to hold, a supplementary condition is needed, such as f injective on an open neighborhood of $\sigma(T)$, or f non-constant on every component of its domain which intersects $\sigma(T)$, [ibid.]. This property has been extended in §.7 to unbounded closed operators. The results of that section are based on [E-W]. The bounded decomposable operator $A = R(\lambda;T)$ in our approach plays an intermediate role. A different approach to the functional calculus for unbounded closed operators is presented in [Su-X.1982]. The bounded version of 7.4 appeared in [Bar-Ka.1973].

The functional calculus, when applicable, provides an easy proof for the spectral duality theorem, which is the topic of the next chapter.

Finally, we mention the papers [Su.1982], [Su.1983] as related to the topic of this chapter.

CHAPTER III. SPECTRAL DUALITY.

The basic theorem of spectral duality asserts that T has the SDP iff T* has the same property, provided that T* is defined. The "only if" part of the theorem, referred to as the *duality theorem*, is presented in §.8, and the "if" part, the *predual theorem*, is the main topic of §.9.

Throughout this chapter, the given closed operator T is assumed to be *densely defined.*

§.8. THE DUALITY THEOREM.

8.1. THEOREM. If T has the SDP then T* has the same property.

PROOF. Given T with the SDP, let $\{G_0, G_1\} \in \operatorname{cov} \sigma(T^*) = \operatorname{cov} \sigma(T)$ with $G_0 \in V_\infty$ and $G_1 \in G^K$. Put $F_1 = G_0^c$, $F_0 = G_1^c$ and note that $X(T, F_0 \cup F_1)$ is a spectral maximal space of T. Then

$$\sigma[T|X(T,F_0 \cup F_1)] \subset [F_0 \cap \sigma(T)] \cup [F_1 \cap \sigma(T)].$$

The closed sets $F_0 \cap \sigma(T)$, $F_1 \cap \sigma(T)$ being disjoint and the latter compact, it follows from 5.12 that

$$X(T,F_0 \cup F_1) = X[T,F_0 \cap \sigma(T)] \oplus \Xi[T,F_1 \cap \sigma(T)].$$

Therefore, every $x \in X(T,F_0 \cup F_1)$ has a unique representation

$$x = x_0 + x_1, \quad x_0 \in X[T,F_0 \cap \sigma(T)], \quad x_1 \in \Xi[T,F_1 \cap \sigma(T)]$$

and there is a constant $M > 0$ with $\|x_0\| + \|x_1\| \le M\|x\|$.

Let $x^* \in X^*$. Define the linear functional x_0^* on $X(T,F_0 \cup F_1)$ by

$$\langle x, x_0^* \rangle = \langle x_0, x^* \rangle. \tag{8.1}$$

Clearly x_0^* is well defined and bounded, as seen from

$$|\langle x, x_0^* \rangle| = |\langle x_0, x^* \rangle| \le \|x_0\| \cdot \|x^*\| \le M\|x\| \cdot \|x^*\|.$$

By the Hahn-Banach theorem, x_0^* can be extended to the entire space X. For $x_1 \in \Xi[T,F_1 \cap \sigma(T)] = \Xi(T,F_1)$, we have

$$\langle x_1, x_0^* \rangle = 0 \tag{8.2}$$

and hence $x_0^* \in \Xi(T,F_1)^a$. Define x_1^* by

$$x^* = x_0^* + x_1^*. \tag{8.3}$$

Note that

$$x_0 \in X[T,F_0 \cap \sigma(T)] = X(T,F_0),$$

use (8.1), (8.2) and obtain succesively

$$\langle x_0, x_1^* \rangle = \langle x_0, x^* \rangle - \langle x_0, x_0^* \rangle = \langle x, x_0^* \rangle$$
$$- \langle x_0, x_0^* \rangle = \langle x_0, x_0^* \rangle - \langle x_0, x_0^* \rangle = 0.$$

Thus $x_1^* \in X(T,F_0)^a$ and hence (8.3) gives rise to the decomposition

(8.4) $$X^* = \Xi(T,F_1)^a + X(T,F_0)^a = Y_0^* + Y_1^*$$

with $Y_0^* = \Xi(T,F_1)^a$, $Y_1^* = X(T,F_0)^a$. By 5.17 (IV,b), $T'' = T/X(T,F_0)$ is bounded and

(8.5) $$\sigma(T'') \subset (\text{Int } F_0)^c \subset \overline{G}_1.$$

To show that the following properties hold

(8.6) (a) $Y_1^* \subset \mathcal{D}_{T^*}$; (b) $Y_1^* \in \text{Inv } T^*$; (c) $T^*|Y_1^* = (T'')^*$,

let $y^* \in Y_1^*$, $x \in \mathcal{D}_T$ and $x'' = x + X(T,F_0) \in X/X(T,F_0)$. For $z \in X(T,F_0)$, one has

(8.7) $$\langle Tx, y^* \rangle = \langle Tx + z, y^* \rangle = \langle (Tx)'', y^* \rangle = \langle T''x'', y^* \rangle.$$

Since T'' is bounded, (8.7) implies

$$|\langle Tx, y^* \rangle| = |\langle T''x'', y^* \rangle| \le \|T''\| \cdot \|x''\| \cdot \|y^*\| \le \|T''\| \cdot \|x\| \cdot \|y^*\|$$

and hence $\langle Tx, y^* \rangle$ is a bounded linear functional of $x \in \mathcal{D}_T$. Therefore, $y^* \in \mathcal{D}_{T^*}$ and hence $Y_1^* \subset \mathcal{D}_{T^*}$. Thus (8.6,a) holds. Let $x_0 \in X(T,F_0) \cap \mathcal{D}_T$. Then $0 = \langle Tx_0, y^* \rangle = \langle x_0, T^*y^* \rangle$. Since $\overline{X(T,F_0) \cap \mathcal{D}_T} = X(T,F_0)$ by 5.15, we have $T^*y^* \in Y_1^*$ and this settles (8.6,b). Finally, (8.7) implies

$$\langle x, T^*y^* \rangle = \langle x'', (T'')^*y^* \rangle$$

or, equivalently,

$$\langle x'', (T^*|Y_1^*)y^* \rangle = \langle x'', (T'')^*y^* \rangle$$

and hence $(T'')^* = T^*|Y_1^*$ which is (8.6,c).

By similar lines, one can show that

(8.8) (a) $Y_0^* \in \text{Inv } T^*$; (b) $(T')^* = T^*|Y_0^*$,

where $T' = T/\Xi(T,F_1)$. By 5.17 (II,b), we have

(8.9) $$\sigma(T') \subset (\text{Int } F_1)^c \subset \overline{G}_0.$$

Now (8.6,c), (8.5), (8.8,b) and (8.9) imply

(8.10) $$\sigma(T^*|Y_1^*) = \sigma(T'') \subset \overline{G}_1; \quad \sigma(T^*|Y_0^*) = \sigma(T') \subset \overline{G}_0.$$

Thus, T* has the SDP, by (8.4), (8.6,a), (8.6,b), (8.8,a) and (8.10). □

§.9. THE PREDUAL THEOREM.

In this section we shall show that T* induces the SDP to T. To set the stage, we shall adopt some preliminary facts from [Bi.1959, pp.382-386]. A couple U_1 and U_2 of an unbounded and a bounded Cauchy domain, related by $U_2 = (\overline{U}_1)^c$, are referred to as *complementary simple sets*. Let W_1 be the set of analytic functions from U_1 to X which vanish at ∞, and W_2 be the set of analytic functions from U_2 to X*. With the help of the seminorms

$$\|f\|_{K_1} = \max\{\|f(\lambda)\| : f \in W_1,\ \lambda \in K_1,\ K_1(\subset U_1) \text{ is compact}\},$$
$$\|g\|_{K_2} = \max\{\|g(\lambda)\| : g \in W_2,\ \lambda \in K_2,\ K_2(\subset U_2) \text{ is compact}\},$$

one can define a locally convex topology on W_1 and W_2, respectively. For i=1,2; let V_i be the subset of W_i on which every function can be extended to be continuous on $\overline{U}_i$. For $f \in V_1$, $g \in V_2$, define

$$\|f\|_{V_1} = \sup\{\|f(\lambda)\| : \lambda \in U_1\},$$
$$\|g\|_{V_2} = \sup\{\|g(\lambda)\| : \lambda \in U_2\},$$

and note that $(V_1, \|\cdot\|_{V_1})$ and $(V_2, \|\cdot\|_{V_2})$ are Banach spaces. For $x \in X$, $\lambda \in U_2$ and $\mu \in U_1$, define

$$\alpha(x,\lambda,\mu) = (\mu-\lambda)^{-1}x.$$

For fixed $x \in X$ and $\lambda \in U_2$, $\alpha(x,\lambda,\cdot)$ is called an *elementary element* of V_1. Let V be the subspace of V_1 spanned by the elementary elements of of V_1. For $f \in V_1$ and $g \in V_2$, with continuous extensions to $C = \partial U_1 = \partial U_2$, the bilinear functional

$$\Phi(f) = \langle f,g\rangle = \frac{1}{2\pi i}\int_C \langle f(\lambda),g(\lambda)\rangle\, d\lambda \tag{9.1}$$

is jointly continuous.

9.1. LEMMA. Let U_1, U_2 be complementary simple sets. With V, V_i and W_i (i=1,2), as defined above, there exists a linear manifold Y in W_2 and a norm on Y such that

(i) Y is a Banach space isometrically isomorphic to V*;

(ii) $V_2 \subset Y$;

(iii) the mappings $V_2 \to Y$ and $Y \to W_2$ are continuous;

(iv) the inner product between V and V_2, defined by (9.1), can be extended to an inner product between V and Y in conjunction with the isometric isomorphism between Y and V*, asserted by (i).

PROOF. (i): Let $\Phi \in V^*$. For $x \in X$ and fixed $\lambda \in U_2$, $\langle\alpha(x,\lambda,\cdot),\Phi\rangle$ is a bounded linear functional on X. Hence there is $g : U_2 \to X^*$ such that

(9.2) $$\langle\alpha(x,\lambda,\cdot),\Phi\rangle = \langle x,g(\lambda)\rangle .$$

It is easily seen that $\lambda \to \langle\alpha(x,\lambda,\cdot),\Phi\rangle$ (and hence $\lambda \to \langle x,g(\lambda)\rangle$) is analytic on U_2. Therefore, g is analytic on U_2.

Let Y be the set of all functions g defined by (9.2). Then $Y \subset W_2$ and the mapping

(9.3) $$\Phi \to g$$

is an isomorphism from V* onto Y. With the norm $\|\cdot\|_Y$, defined by $\|g\|_Y = \|\Phi\|_{V^*}$, $(Y, \|\cdot\|_Y)$ is a Banach space, isometrically isomorphic to V*. Henceforth, we shall identify Y with V*.

(ii): For $g \in V_2$, (9.1) defines a bounded linear functional on V, i.e. $\Phi \in V^*$. For $f = \alpha(x,\lambda,\cdot)$, we have

(9.4) $$\Phi(f) = \frac{1}{2\pi i}\int_C \frac{\langle x,g(\mu)\rangle}{\mu-\lambda}\,d\mu = \langle x,g(\lambda)\rangle.$$

Thus, $g \in V_2$ is the range of the bounded linear functional Φ, and in view of (9.3), $V_2 \subset Y$.

(iii): Let $\{g_n\} \subset V_2$ be a sequence converging to g. The continuity of the mapping $V_2 \to Y$ follows from

$$\|g_n - g\|_Y = \sup\{|\langle f,g_n - g\rangle| : \|f\| = 1,\ f \in V\}$$

$$\leq \frac{1}{2\pi}\int_C |f(\mu)[g_n(\mu) - g(\mu)]|\cdot|d\mu| \leq M\|g_n - g\|_{V_2} \to 0,$$

M being a constant.

Next, we show that the mapping $Y \to W_2$ is continuous. Let $\{g_n\} \subset Y$ be a sequence converging to g. If Φ_n and Φ are the elements in V* which correspond to g_n and g, respectively, then $\|\Phi_n - \Phi\|_{V^*} \to 0$.

For $x \in X$ with $\|x\| \leq 1$, $\lambda \in K = \overline{K} \subset U_2$, it follows from (9.2) that

$$|\langle x, g_n(\lambda) - g(\lambda)\rangle| = |\langle \alpha(x,\lambda,\cdot), \Phi_n - \Phi\rangle|$$
$$\leq \frac{\|x\|}{\delta} \|\Phi_n - \Phi\|_{V^*} \leq \frac{1}{\delta}\|\Phi_n - \Phi\|_{V^*} \to 0,$$

where $\delta = d(K,C)$. Thus, it follows that

$$\sup_{\lambda \in K} \|g_n(\lambda) - g(\lambda)\| \to 0$$

and since the compact $K \subset U_2$ is arbitrary, the sequence $\{g_n\}$ converges to g in the topology of W_2.

(iv): For $f \in V$, $g \in Y$ and $\Phi \in V^*$ corresponding to g, put

$$\langle f,g\rangle = \Phi(f). \tag{9.5}$$

Then, it follows from (9.1), that the inner product between V and Y, as defined by (9.5), is an extension of the inner product between V and V_2. □

The following example shows that, for an open $G \subset \mathbb{C}$, $\overline{X^*(T^*,G)}$ need not be closed in the weak*-topology of X^*, even if T is a bounded decomposable operator.

9.2. EXAMPLE. Let T be the multiplication operator on $X = C[0,1]$, i.e.

$$Tx = tx(t), \quad x \in C[0,1].$$

Then both T and T^* are decomposable. For $F = [0,\frac{1}{2}]$, we have

$$X(T,F) = \{x \in X : \operatorname{supp} x(t) \subset [0,\tfrac{1}{2}]\},$$
$$X(T,F)^a = \{y \in X^* : \operatorname{supp} y(t) \subset [\tfrac{1}{2},1]\},$$

where the dual space X^* is the set of all functions of bounded variation on [0,1], which are continuous from the right on (0,1) and normalized to 0 at $t = 0$. Put

$$y_0(t) = \begin{cases} 0, & \text{if } t \in [0,\frac{1}{2}); \\ 1, & \text{if } t \in [\frac{1}{2},1]. \end{cases}$$

For every $y \in X^*(T^*,F^c)$, we have

$$\|y_0 - y\| = \operatorname{Var}(y_0 - y) \geq 1.$$

Thus

$$d[y_0, X^*(T^*,F^c)] \geq 1, \text{ i.e. } d[y_0, \overline{X^*(T^*,F^c)}] \geq 1,$$

where $\overline{X^*(T^*,F^c)}$ is the closure of $X^*(T^*,F^c)$ in the norm topology of X^*. Thus, it follows that

$$X(T,F)^a \neq \overline{X^*(T^*,F^c)}$$

and hence $\overline{X^*(T^*,F^c)}$ is not closed in the weak*-topology of X^*. □

9.3. THEOREM. If T^* has property (κ) then, for every closed F, $X^*(T^*,F)$ is closed in the weak*-topology of X^*.

PROOF. In view of the Kreĭn-Šmulian theorem, it suffices to show that the norm-closed unit ball S^* of $X^*(T^*,F)$ is closed in the weak*-topology of X^*. Let $\{x^*_\beta\} \subset S^*$ be a net converging to x^* in the weak*-topology of X^*. Let U_1 and U_2 be complementary simple sets with U_1 unbounded and $\overline{U}_2 \cap F = \emptyset$, i.e. $U_1 \supset F$. For each β, the function

$$f^*_\beta(\lambda) = R[\lambda;T^*|X^*(T^*,F)]x^*_\beta$$

is analytic in a neighborhood of $\overline{U}_2$, i.e. $f^*_\beta \in V_2$. Moreover, it follows from

$$\|f^*_\beta(\lambda)\| \leq \|R[\lambda;T^*|X^*(T^*,F)]\| \cdot \|x^*_\beta\| \leq \|R[\lambda;T^*|X^*(T^*,F)]\|$$

and from $\overline{U}_2 \subset \rho[T^*|X^*(T^*,F)]$ that the net $\{f^*_\beta\}$ is bounded in V_2. With Y, as defined in 9.1, the embedding $V_2 \to Y$ being continuous there is $M > 0$ such that

$$\|f^*_\beta\|_Y = \|f^*_\beta\|_{V^*} \leq M\|f^*_\beta\|_{V_2} .$$

Consequently, $\{f^*_\beta\}$ is bounded in V^* and hence $\{f^*_\beta\}$ has a cluster point f^* in the weak*-topology of V^*.

Next, we show that, for every $\lambda \in U_2$, $f^*(\lambda) \in \mathcal{D}_{T^*}$ and

(9.6) $$(\lambda-T^*)f^*(\lambda) = x^*.$$

For $x \in \mathcal{D}_T$ and $g = f^*_\beta$, use (9.2) to write $\langle\alpha(x,\lambda,\cdot),f^*_\beta\rangle = \langle x,f^*_\beta(\lambda)\rangle$;

$$\langle\alpha(Tx,\lambda,\cdot),f^*_\beta\rangle = \langle Tx,f^*_\beta(\lambda)\rangle = \langle x,T^*f^*_\beta(\lambda)\rangle.$$

Consequently, we have

(9.7) $$\langle\alpha(\lambda x-Tx,\lambda,\cdot),f^*_\beta\rangle = \langle(\lambda-T)x,f^*_\beta(\lambda)\rangle = \langle x,(\lambda-T^*)f^*_\beta(\lambda)\rangle = \langle x,x^*_\beta\rangle.$$

Since $x^*_\beta \to x^*$ in the weak*-topology of X^* and f^* is a cluster point of $\{f^*_\beta\}$, (9.7) implies

(9.8) $$\langle\alpha(\lambda x-Tx,\lambda,\cdot),f^*\rangle = \langle x,x^*\rangle.$$

It is easy to see that $\alpha(\lambda x-Tx,\lambda,\cdot)$ is an elementary element of V_1.

Thus quoting (9.2) again, we have

$$<\alpha(\lambda x-Tx,\lambda,\cdot),f^*> = <(\lambda-T)x,f^*(\lambda)>. \tag{9.9}$$

Relation (9.8) coupled with (9.9) gives

$$<(\lambda-T)x,f^*(\lambda)> = <x,x^*>$$

which means that, for $\lambda \in U_2$, $<(\lambda-T)x,f^*(\lambda)>$ is a bounded linear functional of $x \in \mathcal{D}_T$. Consequently, $f^*(\lambda) \in \mathcal{D}_{T^*}$ and (9.6) follows. Thus $U_2 \subset \rho(x^*,T^*)$ and since $U_2 \subset F^c$ is arbitrary, we have $F^c \subset \rho(x^*,T^*)$, i.e. $\sigma(x^*,T^*) \subset F$. Therefore, $x^* \in X^*(T^*,F)$ and since $\|x^*\| \le 1$, we have $x^* \in S^*$. Consequently, $X^*(T^*,F)$ is closed in the weak*-topology of X^*. □

9.4. THEOREM. If T^* has property (κ) then, for every compact F, $\Xi^*(T^*,F)$ is closed in the weak*-topology of X^*.

PROOF. Let Σ^* be the unit ball of $\Xi^*(T^*,F)$ and let the net $\{x^*_\beta\} \subset \Sigma^*$ converge to x^* in the weak*-topology of X^*. It follows from 9.3 that $x^* \in X^*(T^*,F)$. Let $\varepsilon > 0$ be arbitrary. There exists a number $r > \|T^*|\Xi^*(T^*,F)]\|$ such that

$$\|R[\lambda;T^*|\Xi^*(T^*,F)]\| \le \varepsilon \quad \text{for} \quad |\lambda| \ge r. \tag{9.10}$$

Let $R > r$ be arbitrary and consider the bounded Cauchy domain $U_2 = \{\lambda : r < |\lambda| < R\}$. Denote by U_1 the unbounded complementary simple set and put

$$f^*_\beta(\lambda) = R[\lambda;T^*|\Xi^*(T^*,F)]x^*_\beta \quad \text{for} \quad \lambda \in U_2. \tag{9.11}$$

The net $\{f^*_\beta\}$ is bounded in V^* and hence it has, at least, one cluster point f^* in V^*. As in the proof of 9.3, one obtains

$$(\lambda-T^*)f^*(\lambda) = x^* \quad \text{for} \quad \lambda \in U_2.$$

To show that, for $\lambda \in U_2$, one has $\|f^*(\lambda)\| \le \varepsilon$, let $x \in X$ and use (9.2) to write

$$<\alpha(x,\lambda,\cdot),f^*_\beta> = <x,f^*_\beta(\lambda)> , \quad \lambda \in U_2.$$

Then, for $\lambda \in U_2$, with the help of (9.10) and (9.11), one obtains

$$|<\alpha(x,\lambda,\cdot),f^*_\beta>| = |<x,f^*_\beta(\lambda)>| \le \|x\|\cdot\|f^*_\beta(\lambda)\| \le \varepsilon\,\|x\|.$$

Since f^* is a cluster point of $\{f^*_\beta\}$, it follows that

$$|<\alpha(x,\lambda,\cdot),f^*>| \le \varepsilon\,\|x\|, \qquad \lambda \in U_2$$

or, equivalently,

(9.12) $$|\langle x,f^*(\lambda)\rangle| \le \varepsilon \|x\|, \quad \lambda \in U_2.$$

Since $R > r$ is arbitrary, (9.12) implies that

$$\|f^*(\lambda)\| \le \varepsilon \quad \text{for} \quad |\lambda| \ge r.$$

Thus $\lim_{\lambda\to\infty} \|f^*(\lambda)\| = 0$ and therefore $x^* \in \Sigma^*$, by 5.11. □

9.5. THEOREM. Let T^* have the SDP. For a closed $F \in V_\infty$, put $Y = {}^a X^*(T^*,F)]$. Then

$$Y \subset \mathcal{D}_T, \quad Y \in \operatorname{Inv} T, \quad \sigma(T|Y) \subset (\operatorname{Int} F)^c.$$

Moreover, the coinduced $T^{*\wedge} = (T^*)^\wedge = T^*/X^*(T^*,F)$ is the conjugate of $T|Y$.

PROOF. $X^*(T^*,F)$ is closed in the weak*-topology of X^*, by 9.3. Then $Y^a = X^*(T^*,F)$. The spaces Y^* and $X^*/X^*(T^*,F)$ are isometrically isomorphic. In the sequel, we shall identify Y^* with $X^*/X^*(T^*,F)$ and, for an element of $X^*/X^*(T^*,F)$, we write $x^{*\wedge} = (x^*)^\wedge$. Let F_1 be compact such that

(9.13) $$\operatorname{Int} F \cup \operatorname{Int} F_1 = \mathbb{C}.$$

Then, by the SDP of T^*, every $x^* \in X^*$ has a representation

$$x^* = x_0^* + x_1^* \text{ with } x_0^* \in X^*(T^*,F), \quad x_1^* \in \Xi^*(T^*,F_1)$$

and there is a constant $M > 0$ such that $\|x_0^*\| + \|x_1^*\| \le M\|x^*\|$. Since $T^{*\wedge}$ and $[T^*|\Xi^*(T^*,F_1)]/X^*(T^*,F) \cap \Xi^*(T^*,F_1)$ are similar, $T^{*\wedge}$ is bounded and $\sigma(T^{*\wedge}) \subset F_1$. Since F_1, subject to (9.13), is arbitrary, we have

(9.14) $$\sigma(T^{*\wedge}) \subset (\operatorname{Int} F)^c.$$

Next, let $\{x_\beta^{*\wedge}\}$ be a bounded net in $X^*/X^*(T^*,F)$ converging to $x^{*\wedge}$ in the weak*-topology of Y^*, i.e. for every $x \in Y$,

(9.15) $$\lim_\beta \langle x,x_\beta^{*\wedge}\rangle = \langle x,x^{*\wedge}\rangle.$$

For every $x_\beta^{*\wedge}$ there is $x_\beta^* \in x_\beta^{*\wedge}$ with $\|x_\beta^*\| \le 2\|x_\beta^{*\wedge}\|$ and, for each β, we have a representation

$$x_\beta^* = x_{\beta 0}^* + x_{\beta 1}^*, \quad x_{\beta 0}^* \in X^*(T^*,F), \quad x_{\beta 1}^* \in \Xi^*(T^*,F_1),$$

with

$$\|x_{\beta 0}^*\| + \|x_{\beta 1}^*\| \le M\|x_\beta^*\|.$$

Since $\{x_\beta^*\}$ is bounded, so is $\{x_{\beta 1}^*\}$ and hence the latter has a cluster point $y^* \in \Xi^*(T,F_1)$ in the weak*-topology of X^*, by 9.4. Consequently, for every $x \in \mathcal{D}_T$, $\varepsilon > 0$ and any index γ, there is an index $\beta > \gamma$ such that $|\langle Tx, x^*_{\beta 1} - y^*\rangle| < \varepsilon$ or, equivalently,

$$|\langle x, T^*(x^*_{\beta 1} - y^*)\rangle| < \varepsilon, \quad \text{for some } \beta > \gamma \text{ depending on } x. \tag{9.16}$$

Since T is densely defined, $T^*|\Xi^*(T^*,F_1)$ is bounded and $\{x^*_{\beta 1}\}$ is a bounded net, it follows that (9.16) holds for each $x \in X$. Evidently, $y^{*\wedge}$ is a cluster point of $\{x^{*\wedge}_{\beta 1} = x^{*\wedge}_{\beta}\}$. It follows from (9.15) that $y^{*\wedge} = x^{*\wedge}$ and hence (9.16) gives

$$|\langle x, T^{*\wedge}(x^{*\wedge}_{\beta} - x^{*\wedge})\rangle| < \varepsilon, \quad x \in Y, \quad \text{for some } \beta > \gamma. \tag{9.17}$$

Note that, for $x \in Y$, $f(x^{*\wedge}) = \langle x, T^{*\wedge}x^{*\wedge}\rangle$ is a bounded linear functional on $X^*/X^*(T^*,F)$. More than this is true: f is continuous in the *bounded weak*-topology*, i.e. f is continuous in the BY-*topology* (for the BX-topology see e.g. [Du-S.1967; Definition V. 5.3, p.427]). To see that f is continuous in the bounded weak*-topology, it suffices to show that, for every closed δ, the set

$$\{x^{*\wedge} : f(x^{*\wedge}) \in \delta\} \subset X^*/X^*(T^*,F) \tag{9.18}$$

is closed in the BY-topology. Accordingly, we have to show that, for every $a > 0$, the intersection

$$\{x^{*\wedge} : f(x^{*\wedge}) \in \delta\} \cap aS^{*\wedge} \tag{9.19}$$

is closed in the weak*-topology of $X^*/X^*(T^*,F)$, where $S^{*\wedge}$ is the closed unit ball of $X^*/X^*(T^*,F)$. Let $\{x^{*\wedge}_{\beta}\} \subset \{x^{*\wedge} : f(x^{*\wedge}) \in \delta\} \cap aS^{*\wedge}$ be a net converging to $\xi^{*\wedge}$ in the weak*-topology of $X^*/X^*(T^*,F)$. It follows from (9.17) that

$$|f(x^{*\wedge}_{\beta}) - f(\xi^{*\wedge})| < \varepsilon \quad \text{for some } \beta > \gamma.$$

Since $f(x^{*\wedge}_{\beta}) \in \delta$, we have $f(\xi^{*\wedge}) \in \delta$. Evidently, $\|\xi^{*\wedge}\| \le a$. Thus, it follows that $\xi^{*\wedge} \in \{x^{*\wedge} : f(x^{*\wedge}) \in \delta\} \cap aS^*$ and hence the set (9.19) is closed in the BY-topology. Equivalently, (9.18) is closed in the weak*-topology of $X^*/X^*(T^*,F)$. By a known property (e.g.[Du-S.1967; V. 5.6]), f is continuous in the weak*-topology of $X^*/X^*(T^*,F)$ and hence there is a unique $y \in Y$ such that $f(x^{*\wedge}) = \langle y, x^{*\wedge}\rangle$ or, equivalently,

$$\langle x, T^{*\wedge}x^{*\wedge}\rangle = \langle y, x^{*\wedge}\rangle. \tag{9.20}$$

Put $Ax = y$ and note that A is a linear operator, well-defined on Y. Moreover, (9.20) implies $\|Ax\| \le \|T^{*\wedge}\| \cdot \|x\|$, so $A \in B(Y)$ and $A^* = T^{*\wedge}$.

Finally, we show that $Y \in \text{Inv } T$ and $A = T|Y$. With the mapping $V : X \oplus X \to X \oplus X$, defined by $V(x,y) = (-y,x)$, and the graph $G(\cdot)$, (9.20) implies

$$V(x,Ax) \perp G(T^{*\wedge}).$$

For $z_1, z_2 \in X^*(T^*,F)$, one has

$$\langle V(x,Ax),(x^*,T^*x^*)\rangle = \langle(-Ax,x),(x^*+z_1,T^*x^*+z_2)\rangle$$
$$= \langle(-Ax,x),(x^{*\wedge},T^{*\wedge}x^{*\wedge})\rangle = 0.$$

Thus $V(x,Ax) \subset {}^aG(T^*) = VG(T)$ and hence $(x,Ax) \in G(T)$. Therefore, $x \in \mathcal{D}_T$, $A = T|Y$, $Y \subset \mathcal{D}_T$, $Y \in \text{Inv } T$ and $(T|Y)^* = T^{*\wedge}$. These relations together with (9.14) conclude the proof. □

9.6. THE PREDUAL THEOREM. T has the SDP if T* has the same property.

PROOF. Assume that T* has the SDP and let $\{G_0,G_1\} \in \text{cov } \mathbb{C}$ with $G_0 \in V_\infty$ and $G_1 \in G^K$. Then $F_0 = G_1^c \in V_\infty$, $F_1 = G_0^c$ is compact and $F_0 \cap F_1 = \emptyset$. It follows from 9.5 that the subspace $Y_1 = {}^aX^*(T^*,F_0)$ has the following properties:

$$Y_1 \subset \mathcal{D}_T,\quad Y_1 \in \text{Inv } T \text{ and } \sigma(T|Y_1) \subset \overline{G}_1. \tag{9.21}$$

Next, consider the subspace $Y_0 = {}^a\Xi^*(T^*,F_1)$. For $x \in Y_0 \cap \mathcal{D}_T$, $x^* \in \Xi^*(T^*,F_1)$, we have

$$\langle Tx,x^*\rangle = \langle x,T^*x^*\rangle = 0$$

and hence $Tx \in Y_0$, implying that $Y_0 \in \text{Inv } T$. The subspace

$$Z^* = X^*(T^*,F_0) \oplus \Xi^*(T^*,F_1) \tag{9.22}$$

is closed in X*. Moreover, since $X^*(T^*,F_0)$ and $\Xi^*(T^*,F_1)$ are closed in the weak*-topology of X*, by 9.3 and 9.4, so is Z*.

Let $x \in X$, $x^* \in Z^*$ and x_0^* be the projection of x^* onto $X^*(T^*,F_0)$, in conjunction with (9.22). The linear functional x_0 on Z*, defined by

$$\langle x_0,x^*\rangle = \langle x,x_0^*\rangle, \tag{9.23}$$

is continuous in the weak*-topology. By the Hahn-Banach theorem on locally convex spaces, x_0 can be extended to a linear functional on X*, which is continuous in the weak*-topology. Therefore, $x_0 \in X$. Since the projection x_0^* of $x^* \in \Xi^*(T^*,F_1)$ onto $X^*(T^*,F_0)$ is zero, it follows from (9.23) that

$$\langle x_0, x^* \rangle = 0, \quad x^* \in \Xi^*(T, F_1)$$

and hence $x_0 \in {}^a\Xi^*(T^*, F_1) = Y_0$. Let $x_1 = x - x_0$. Then, for $x^* \in X^*(T^*, F_0)$, (9.23) shows that $\langle x_1, x^* \rangle = 0$ and hence $x_1 \in {}^aX^*(T^*, F_0) = Y_1$. Since $x \in X$ is arbitrary, the representation

$$x = x_0 + x_1 \quad \text{with} \quad x_i \in Y_i \quad (i=0,1)$$

implies

$$X = Y_0 + Y_1. \tag{9.24}$$

By 5.15, $T|Y_0$ is densely defined and therefore $(T|Y_0)^*$ exists. It follows easily that $(T|Y_0)^*$ is unitarily equivalent to $T^*/\Xi^*(T^*, F_1)$. In view of 5.17, one obtains

$$\sigma(T|Y_0) = \sigma[T^*/\Xi^*(T^*, F_1)] \subset (\text{Int } F_1)^c \subset \overline{G}_0. \tag{9.25}$$

In conclusion, T has the SDP by (9.24), (9.21) and (9.25). □

If T has property (κ) then, for each compact F, $\Xi(T,F)$ is the T-bounded spectral maximal space associated to $X(T,F)$. Next, we extend the definition of $\Xi(T,\cdot)$ to the family G of all open subsets of $\mathbb{C}$.

9.7. DEFINITION. If T has the SVEP then, for each $G \in G$,

$$\Xi(T,G) = \bigcup \{\Xi(T,F) : F \text{ is compact and } F \subset G\}. \tag{9.26}$$

Evidently, $\Xi(T,G)$ is a linear manifold in X, invariant under T. As regarding $\overline{X(T,G)}$ and $\overline{\Xi(T,G)}$, they have the following properties in duality:

9.8. THEOREM. Let T or T* have the SDP.

(I) For every open $G \in V_\infty$, $Y = \overline{X(T,G)}$ has the properties:

(i) $Y \in \text{Inv } T$, $\sigma(T|Y) \subset \overline{G}$;

(ii) $\hat{T} = T/Y$ is bounded and $Y^a = \Xi^*(T^*, G^c)$.

(II) For every $G \in G^K$, $Z = \overline{\Xi(T,G)}$ has the properties:

(i) $Z \in \text{Inv } T$, $\sigma(T|Z) \subset \overline{G}$;

(ii) $\tilde{T} = T/Z$ is closed and $Z^a = X^*(T^*, G^c)$.

PROOF. We confine the proof to (I), that of (II) is similar. If T or T* has the SDP then both have the same property. Given an open $G \in V_\infty$, let $Y = \overline{X(T,G)}$ and denote $F = G^c$. For $x \in X(T,G)$ and $x^* \in \Xi^*(T,F)$, define

$$f(\lambda) = \begin{cases} \langle x(\lambda),x^*\rangle, & \text{if } \lambda \in \rho(x,T); \\ \langle x,x^*(\lambda)\rangle, & \text{if } \lambda \in \rho(x^*,T^*). \end{cases}$$

It is transparent that f is a well-defined entire function. For $C = \{\lambda : |\lambda| = \|T^*|\Xi^*(T^*,F)\| + 1\}$, one obtains

$$\langle x,x^*\rangle = \int_C \langle x,x^*(\lambda)\rangle\, d\lambda = \int_C f(\lambda)d\lambda = 0$$

and hence $x^* \in Y^a$. Therefore, $Y^a \supset \Xi^*(T^*,F)$. To obtain the opposite inclusion, let K be compact with $\operatorname{Int} K \supset F$. It follows from the proof of 9.6 that $Y_K = {}^a\Xi^*(T^*,K)$ is invariant under T and

(9.27) $$\sigma(T|Y_K) = \sigma[T^*/\Xi^*(T^*,K)] \subset (\operatorname{Int} K)^c \subset F^c = G.$$

Thus $Y_K \subset Y$, by (9.27). It follows from the equalities

$$[\vee\{Y_K : \operatorname{Int} K \supset F,\ K \text{ is compact}\}]^a$$
$$= \bigcap\{Y_K^a : \operatorname{Int} K \supset F,\ K \text{ is compact}\}$$
$$= \bigcap\{\Xi^*(T^*,K) : \operatorname{Int} K \supset F,\ K \text{ is compact}\} = \Xi^*(T^*,F)$$

and from the inclusion

$$\vee\{Y_K : \operatorname{Int} K \supset F,\ K \text{ is compact}\} \subset Y,$$

that $Y^a \subset \Xi^*(T^*,F)$. Thus $Y^a = \Xi^*(T^*,F)$, i.e. $X(T,G)^a = \Xi^*(T^*,F)$. F being compact, ${}^a\Xi^*(T^*,F)$ is invariant under T or, equivalently,

$$Y = \overline{X(T,G)} \in \operatorname{Inv} T.$$

To conclude the proof, note that

$$\sigma[T|\overline{X(T,G)}] = \sigma[T|{}^a\Xi^*(T^*,F)]$$
$$= \sigma[T^*/\Xi^*(T^*,F)] \subset (\operatorname{Int} F)^c \subset \overline{G}$$

and $T/\overline{X(T,G)}$ is bounded by 3.4. □

9.9. REMARK. The assertions

$$Z \in \operatorname{Inv} T,\quad \sigma(T|Z) \subset \overline{G} \quad \text{and} \quad \hat{T} = T/Z \text{ is closed}$$

in 9.8 (II) hold, for any closed T which has the SDP, without the domain-density assumption on T.

For bounded operators, the *bipolar relation* between the spectral manifold and its annihilator assumes the following form.

9.10. THEOREM. Given $T \in B(X)$, suppose that T or T^* is decomposable. Then, for any closed F,

$$X(T,F) = {}^aX^*(T^*,F^c) \tag{9.28}$$

or, by the bipolar theorem,

$$X(T,F)^a = \overline{X^*(T^*,F^c)}^w .$$

PROOF. If T or T^* is decomposable then both are. Given a closed F, denote $G_0 = F^c$. Choose $G_1 \in G$ to satisfy

$$\{G_0,G_1\} \in \text{cov}\, \sigma(T^*). \tag{9.29}$$

First, we obtain the decomposition

$$X^* = \overline{X^*(T^*,G_0)} + X^*(T^*,\overline{G}_1). \tag{9.30}$$

Indeed, it suffices to select an open H_0 such that $\overline{H}_0 \subset G_0$ and $\{H_0,G_1\} \in \text{cov}\, \sigma(T^*)$. Then (9.30) follows from $\overline{X^*(T^*,G_0)} \supset X^*(T^*,\overline{H}_0)$ and

$$X^* = X^*(T^*,\overline{H}_0) + X^*(T^*,\overline{G}_1).$$

Now, 3.4 and (9.30) imply $\sigma[(T^*)\hat{}\,] \subset \overline{G}_1$ and since G_1 is an arbitrary member of the cover (9.29), one has

$$\sigma[(T^*)\hat{}\,] \subset {G_0}^c = F.$$

Therefore, the spectrum of the restriction $T|{}^aX^*(T^*,F^c)$ satisfies

$$\sigma[T|{}^aX^*(T^*,F^c)] \subset F$$

and this leads one to

$$X(T,F) \supset {}^aX^*(T^*,F^c).$$

The opposite inclusion being evident, (9.28) follows. □

§.10. THE ANALYTICALLY INVARIANT SUBSPACE IN DUALITY.

A basic property of analytically invariant subspaces, stated by 4.14, can be retrieved in duality.

We denote by J and K the canonical embeddings of X into X^{**} and of X^* into X^{***}, respectively. The definitions of J and K are

$$\langle x^*,Jx\rangle = \langle x,x^*\rangle \quad \text{and} \quad \langle x^{**},Kx^*\rangle = \langle x^*,x^{**}\rangle.$$

10.1. LEMMA. The following direct sum decomposition holds

(10.1) $$X^{***} = KX^* \oplus (JX)^a.$$

PROOF. For every $x^{***} \in X^{***}$, there is a unique $x^* \in X^*$ such that

$$2\pi i\langle x,x^*\rangle = \langle Jx,x^{***}\rangle, \quad x \in X.$$

Then, it follows from the equalities

$$\langle Jx,Kx^*\rangle = \langle x^*,Jx\rangle = \langle x,x^*\rangle = \langle Jx,x^{***}\rangle$$

that

$$x^{***} - Kx^* \in (Jx)^a.$$

Thus, there is $y^{***} \in (JX)^a$ such that x^{***} has a representation

$$x^{***} = Kx^* + y^{***}$$

and since $KX^* \cap (JX)^a = \{0\}$, the decomposition (10.1) is obtained. □

Our study proceeds through the successive conjugates of the given densely defined closed operator T. The domain density conditions:

(*) T^* is densely defined;

(**) T^* and T^{**} are densely defined;

(***) T^*, T^{**} and T^{***} are densely defined;

(****) T^*, T^{**}, T^{***} and T^{****} are densely defined;

form the list of the hypotheses. In view of (10.1), we shall henceforth denote by P the projection of X^{***} onto KX^* along $(JX)^a$.

10.2. THEOREM. Given T, the following properties hold.

(I) If the density condition (*) is satisfied then, for every $x \in \mathcal{D}_T$, we have $Jx \in \mathcal{D}_{T^{**}}$ and $T^{**}Jx = JTx$; likewise

(I') if the density condition (**) is satisfied then, for each $x^* \in \mathcal{D}_{T^*}$, we have $Kx^* \in \mathcal{D}_{T^{***}}$ and $T^{***}Kx^* = KT^*x^*$.

(II) Assume that (**) holds and $x^{***} \in KX^*$. If $\langle T^{**}Jx,x^{***}\rangle$ is a bounded linear functional of $Jx \in J\mathcal{D}_T$, then $x^{***} \in K\mathcal{D}_{T^*}$ and $T^{***}x^{***} = KT^*K^{-1}x^{***}$.

PROOF. (I): It follows from the graph $G(\cdot)$ and the inverse graph $VG(\cdot)$ related identities

$$G(T) = {}^a[VG(T^*)], \quad [VG(T^*)]^a = G(T^{**})$$

that, for every $(x,Tx) \in G(T)$, we have $(Jx,JTx) \in G(T^{**})$. Equivalently, $Jx \in \mathcal{D}_{T^{**}}$ and $T^{**}Jx = JTx$.

(I'): follows directly from (I) with the original space X replaced by X* and the embedding J replaced by K.

(II): Let $x \in \mathcal{D}_T$ and suppose that $\langle T^{**}Jx, x^{***}\rangle$ is a bounded linear functional of Jx. With the help of (I), we obtain

$$\langle Tx, K^{-1}x^{***}\rangle = \langle K^{-1}x^{***}, JTx\rangle = \langle K^{-1}x^{***}, T^{**}Jx\rangle = \langle T^{**}Jx, x^{***}\rangle$$

and hence $\langle Tx, K^{-1}x^{***}\rangle$ is a bounded linear functional of x. Then $K^{-1}x^{***} \in \mathcal{D}_{T^*}$ and, by using (I'), we obtain $x^{***} \in \mathcal{D}_{T^{***}}$ and $T^{***}x^{***} = KT^*K^{-1}x^{***}$. □

10.3. COROLLARY. Given T, assume that (**) holds. Then

$$K\mathcal{D}_{T^*} = KX^* \cap \mathcal{D}_{T^{***}} .$$

PROOF. It follows from 10.2 (I') that

$$K\mathcal{D}_{T^*} \subset KX^* \cap \mathcal{D}_{T^{***}} .$$

For $x^{***} \in KX^* \cap \mathcal{D}_{T^{***}}$, $\langle T^{**}Jx, x^{***}\rangle = \langle Jx, T^{***}x^{***}\rangle$ is a bounded linear functional of Jx. Then, by 10.2 (II), $x^{***} \in K\mathcal{D}_{T^*}$ and hence

$$KX^* \cap \mathcal{D}_{T^{***}} \subset K\mathcal{D}_{T^*} . \quad \square$$

10.4. LEMMA. Given T, assume that (**) holds. Then, the projection P commutes with T***.

PROOF. Let $x^{***} \in \mathcal{D}_{T^{***}}$. Then $Px^{***} \in KX^*$. For $x \in \mathcal{D}_T$, we have

$$\langle T^{**}Jx, Px^{***}\rangle = \langle T^{**}Jx, Px^{***}\rangle + \langle T^{**}Jx, (I-P)x^{***}\rangle$$

$$= \langle T^{**}Jx, x^{***}\rangle = \langle Jx, T^{***}x^{***}\rangle = \langle Jx, PT^{***}x^{***}\rangle$$

and hence $\langle T^{**}Jx, Px^{***}\rangle$ is a bounded linear functional of $Jx \in J\mathcal{D}_T$. By 10.2 (II), $Px^{***} \in \mathcal{D}_{T^{***}}$. Consequently,

$$\langle T^{**}Jx, Px^{***}\rangle = \langle Jx, T^{***}Px^{***}\rangle$$

implies

$$T^{***}Px^{***} = PT^{***}x^{***} . \quad \square$$

10.5. COROLLARY. Given T, assume that (***) holds. Then the projection P* commutes with T****.

10.6. THEOREM. Given T, assume that (***) holds. Then

$$J\mathcal{D}_T = JX \cap \mathcal{D}_{T^{**}}.$$

Moreover, $T^{***}|(JX)^a$ is densely defined and, for every $Jx \in J\mathcal{D}_T$, we have

$$JTx = T^{**}Jx.$$

PROOF. Inclusion

$$J\mathcal{D}_T \subset JX \cap \mathcal{D}_{T^{**}}$$

follows from 10.2 (I). Let $x^{***} \in (JX)^a$. There is a sequence $\{x_n^{***}\} \subset \mathcal{D}_{T^{***}}$ convergent to x^{***}. In view of 10.4, for every n, $Px_n^{***} \in \mathcal{D}_{T^{***}}$ and hence $(I-P)x_n^{***} \in \mathcal{D}_{T^{***}}$. Thus, as $n \to \infty$,

$$(I-P)x_n^{***} \to (I-P)x^{***} = x^{***}$$

and hence $(JX)^a \cap \mathcal{D}_{T^{***}}$ is dense in $(JX)^a$, i.e. $T^{***}|(JX)^a$ is densely defined. Let $Jx \in JX \cap \mathcal{D}_{T^{**}}$. For every $x^{***} \in (JX)^a \cap \mathcal{D}_{T^{***}}$, we have

(10.2) $$0 = \langle Jx, T^{***}x^{***}\rangle = \langle T^{**}Jx, x^{***}\rangle.$$

Since $(JX)^a \cap \mathcal{D}_{T^{***}}$ is dense in $(JX)^a$, it follows from (10.2) that $T^{**}Jx \in JX$. For $x^* \in \mathcal{D}_{T^*}$, 10.2 (I') gives $T^{***}Kx^* = KT^*x^*$ and hence we obtain

$$\langle x, T^*x^*\rangle = \langle Jx, KT^*x^*\rangle = \langle Jx, T^{***}Kx^*\rangle$$

$$= \langle T^{**}Jx, Kx^*\rangle = \langle x^*, T^{**}Jx\rangle = \langle J^{-1}T^{**}Jx, x^*\rangle.$$

This means that $(x, J^{-1}T^{**}Jx) \in {}^a[VG(T^*)] = G(T)$. Consequently, $x \in \mathcal{D}_T$ and $J^{-1}T^{**}Jx = Tx$, i.e.

$$T^{**}Jx = JTx.$$

Since Jx is arbitrary in $JX \cap \mathcal{D}_{T^{**}}$, it follows that $JX \cap \mathcal{D}_{T^{**}} \subset J\mathcal{D}_T$. □

10.7. THEOREM. Given T, assume that (***) holds. If T^{****} has the SVEP then JX is analytically invariant under T^{**}.

PROOF. We let the proof run through three parts.

Part I. In this part, we show that $(X^{**}/JX)^{**}$ and $N(P^*)$ are topologically isomorphic, where $N(\cdot)$ denotes the null space.

In view of 10.1, we have

(10.3) $$(X^{**}/JX)^* = (JX)^a = N(P);$$

(10.4) $$(X^{***}/KX^*)^* = (KX^*)^a = N(P^*).$$

Since, evidently, X^{***}/KX^* and $N(P)$ are topologically isomorphic, it follows from (10.3), (10.4) that $(X^{**}/JX)^{**}$ and $N(P^*)$ are topologically isomorphic.

Part II. In this intermediary part, we show that T^{**}/JX is closable and $(T^{**}/JX)^{**}$ is similar to $T^{****}|N(P^*)$.

In view of 10.6, JX is invariant under T^{**} and hence we can define T^{**}/JX. Quoting again 10.6, $T^{***}|N(P)$ is densely defined and then 3.5 implies that T^{**}/JX is closable and $(T^{**}/JX)^* = T^{***}|N(P)$.

Next, we show that $(T^{***})^\wedge = T^{***}/KX^*$ and $V^{***} = T^{***}|N(P)$ are similar. Let $A : X^{***}/KX^* \to N(P)$ be the topological isomorphism $A(x^{***})^\wedge = (I-P)x^{***}$, where $(x^{***})^\wedge \in X^{***}/KX^*$ corresponds to $x^{***} \in X^{***}$. Let $x^{***} \in \mathcal{D}_{T^{***}}$. Then $(x^{***})^\wedge \in \mathcal{D}_{(T^{***})^\wedge}$ and it follows from 10.4 that

$$(I-P)x^{***} \in \mathcal{D}_{V^{***}}\,.$$

Since

$$A(x^{***})^\wedge = (I-P)x^{***},$$

we have

$$A\mathcal{D}_{(T^{***})^\wedge} \subset \mathcal{D}_{V^{***}}\,.$$

Conversely, let $x^{***} \in \mathcal{D}_{V^{***}}$. Then $x^{***} \in \mathcal{D}_{T^{***}} \cap N(P)$, furthermore, $(x^{***})^\wedge \in \mathcal{D}_{(T^{***})^\wedge}$ and $A(x^{***})^\wedge = x^{***}$. Consequently, we have

$$A\mathcal{D}_{(T^{***})^\wedge} = \mathcal{D}_{V^{***}}\,. \tag{10.5}$$

Now, letting $x^{***} \in \mathcal{D}_{T^{***}}$, it follows from (10.5) and from the equalities

$$A(T^{***})^\wedge(x^{***})^\wedge = A(T^{***}x^{***})^\wedge = (I-P)T^{***}x^{***}$$
$$= T^{***}(I-P)x^{***} = V^{***}A(x^{***})^\wedge,$$

that $(T^{***})^\wedge$ and V^{***} are similar. Therefore, $(T^{***})^\wedge$ is closed. Since, in view of (10.4), we have

$$[(T^{***})^\wedge]^* = T^{****}|N(P^*),$$

it follows that $[T^{***}|N(P)]^*$ and $T^{****}|N(P^*)$ are similar. Therefore, $(T^{**}/JX)^{**}$ and $T^{****}|N(P^*)$ are similar.

Part III. In this final part, the conclusion $JX \in AI(T^{**})$ will be reached. Let $f^{**} : \omega \to \mathcal{D}_{T^{**}}$ be analytic and satisfy condition

$$(\lambda - T^{**})f^{**}(\lambda) \in JX \quad \text{on an open connected } \omega. \tag{10.6}$$

On the quotient space X^{**}/JX, with $(f^{**})^\wedge$ corresponding to f^{**}, (10.6) gives rise to the equation

(10.7) $$(\lambda - T^{**}/JX)(f^{**})^{\wedge}(\lambda) = \hat{0} \text{ on } \omega.$$

Since, by hypothesis, $T^{****}|N(P^*)$ has the SVEP, $(T^{**}/JX)^{**}$ has the same property. Therefore, the minimal closed extension $\overline{T^{**}/JX}$ of T^{**}/JX has the SVEP and (10.7) implies that $(f^{**})^{\wedge}(\lambda) = \hat{0}$ or, equivalently, $f^{**}(\lambda) \in JX$ on ω. □

10.8. COROLLARY. Given $T \in B(X)$, if T^{****} has the SVEP then JX is analytically invariant under T^{**}.

10.9. COROLLARY. Given T, assume that (***) holds. If T^{****} has the SVEP then, for any $E \subset \mathbb{C}$, we have

(10.8) $$JX(T,E) = X^{**}(T^{**},E) \cap JX.$$

PROOF. It follows from the hypothesis on T^{****} that T^{**} as well as T have the SVEP. Thus, the spectral manifolds $X^{**}(T^{**},E)$ and $X(T,E)$ are defined. The inclusion

(10.9) $$JX(T,E) \subset X^{**}(T^{**},E) \cap JX$$

is evident. To obtain the opposite inclusion, let $Jx \in X^{**}(T^{**},E) \cap JX$. The resolvent function $x^{**}(\cdot)$ of Jx identically verifies equation

$$(\lambda - T^{**})x^{**}(\lambda) = Jx \text{ on } \rho(Jx,T^{**}).$$

Since, by 10.7, $JX \in AI(T^{**})$, for every $\lambda \in \rho(Jx,T^{**})$, we have $x^{**}(\lambda) \in JX$. Let $f(\lambda) = J^{-1}x^{**}(\lambda)$ for $\lambda \in \rho(Jx,T^{**})$. Then $f : \rho(Jx,T^{**}) \to \mathcal{D}_T$ is analytic on $\rho(Jx,T^{**})$ and

$$(\lambda - T)f(\lambda) = (\lambda - T)J^{-1}x^{**}(\lambda) = J^{-1}(\lambda - T^{**})x^{**}(\lambda) = J^{-1}Jx = x.$$

Thus, $\rho(x,T) \supset \rho(Jx,T^{**})$ or, equivalently,

$$\sigma(x,T) \subset \sigma(Jx,T^{**}) \subset E$$

and hence $x \in X(T,E)$. The opposite of inclusion (10.9) follows and this concludes the proof. □

A property derived from the SVEP of the fifth conjugate will later be applied in the theory of strongly decomposable operators.

10.10. COROLLARY, Given T, assume that (****) holds. If T^{*****} has the SVEP then, for any $E \subset \mathbb{C}$,

(10.10) $$KX^*(T^*,E) = PX^{***}(T^{***},E).$$

PROOF. The inclusion

(10.11) $$KX^*(T^*,E) \subset PX^{***}(T^{***},E)$$

is evident. To obtain the opposite inclusion, let $x^{***} \in X^{***}(T^{***},E)$. Since P commutes with T^{***}, $\sigma(Px^{***},T^{***}) \subset E$ and hence

$$Px^{***} \in X^{***}(T^{***},E) \cap KX^*.$$

Now (10.8) applied to T^* gives

$$KX^*(T^*,E) = X^{***}(T^{***},E) \cap KX^*.$$

Thus $Px^{***} \in KX^*(T^*,E)$ and the opposite of (10.11) follows. □

NOTES AND COMMENTS

The importance of the spectral duality problem cannot be overemphasized. Although it has received considerable attention during the last fifteen years, quite recently there were many loose ends and unsolved problems.

With the consistent usage of the terminology in this chapter, we call a duality theorem the event when a certain type of spectral decomposition of the underlying space X by T is carried over to a spectral decomposition of the dual space X^* by T^*. The reciprocal phenomenon: the spectral decomposition of X^* by T^* induces a spectral decomposition of X by the original operator T, is referred to as a predual theorem.

On the duality theorem for bounded operators, Frunză ([Fr.1971] and [Fr.1976]) proved that if T is decomposable on X then T^* is decomposable on X^*. Vasilescu extended the duality theorem to (S,1)-decomposable operators [V.1971,a] which, by virtue of the equivalence of the bounded (S,1)- and S- decomposable operators [N.1979], further extends the duality theorem to S-decomposable operators (see also [V.1982, IV. Theorem 5.5]).

An attempt to prove the predual theorem was made in [Fr.1983]. Unfortunately, the author incorrectly asserted that

$$(\dagger) \qquad X(T,K)^a = \overline{X^*(T^*,K^c)}, \quad \text{for } K \text{ compact},$$

with the closure in the right-hand side taken in the weak-topology $\tau(X^*,X^{**})$, whereas the correct bipolar relation is

$$X(T,K)^a = \overline{X^*(T^*,K^c)}^{w},$$

with the closure in the weak*-topology $\tau(X^*,X)$, as shown by 9.2 and 9.10 in this text. The error (†) invalidates the proof of the predual theorem, as outlined in [W.1984,a].

The predual theorem for bounded S-decomposable operators was proved in [W-Li.1984]. The first proof of the predual theorem for unbounded closed operators appeared in [E-W.1984,a], under the additional domain-density condition (***), (listed in §.10 following the proof of 10.1). Specifically, the proofs of 9.3 and 9.4 used the domain-density condition (**), but for the proof of a key property to the predual theorem (partly contained in 9.5), (***) was the basic hypothesis. In a subsequent paper by the authors (Appendix A, and referred to in [W-E.1983]), the domain density condition of T*** could be deleted. Finally, in the version presented in §.9 of this chapter, the authors could rescind the last constraint (**), [W-E,a].

For a bounded operator, the proofs of the duality and predual theorems are much simpler. In fact, if we assume that $T \in B(X)$ has the SDP then, for any $G \in G^K$, the subspace $Y^* = \dot{X}(T,\overline{G})^a$ satisfies the following conditions:

$$Y^* \in \text{Inv } T^*, \quad \sigma(T^*|Y^*) = \sigma[T/X(T,\overline{G})] \subset (\overline{G})^c \subset G^c,$$

$$\sigma(T^*/Y^*) = \sigma[T|X(T,\overline{G})] \subset \overline{G}.$$

Subsequently, T* has the SDP, by 5.17 (VI). If T* is assumed to have the SDP, then we can substantially simplify the proof of 9.3 and delete 9.4.

Under the extra condition: $\rho(T) \neq \emptyset$, the functional calculus can be used for the proof of the spectral duality theorem. In fact, if we assume that a closed operator T, with $\rho(T) \neq \emptyset$ is densely defined then, by 7.2 and the spectral duality theorem for bounded decomposable operators, the spectral duality theorem follows through the following equivalent statements: T has the SDP iff A $(= R(\alpha;T),\ \alpha \in \rho(T))$ is decomposable iff A* is decomposable iff T* has the SDP.

Lemma 10.1 appeared in [W-Li.1984], 10.2-10.6 are included in [E-W.1984,a], 10.7 is in [W-Li.1984] and [E-W].

A recent work on the duality problem for unbounded decomposable operators is [Su.1984,a].

CHAPTER IV. SPECTRAL RESOLVENTS.

Glancing back at the constructs of the spectral decomposition problem, we see that they consist of two kinds of objects: the open sets of the complex plane and the invariant subspaces of X (under the given closed operator T). The correlation between these two structures, in the framework of a spectral decomposition, is the *spectral resolvent* concept.

In this chapter, we adopt a few supplementary notations. Occasionally, we used the notation G for the family of all open sets in $\mathbb{C}$. The counterpart F will denote the collection of all closed subsets of $\mathbb{C}$. As a subcollection, we write F^K for the family of all compact sets in $\mathbb{C}$. $S(X)$ will denote the lattice of all subspaces (closed linear manifolds) of X.

§.11. SPECTRAL RESOLVENTS. GENERAL PROPERTIES.

11.1. DEFINITION. Given T, a map $E : G \to \text{Inv } T$ is called a spectral resolvent of T if the following conditions are satisfied:

(I) $E(G) \subset \mathcal{D}_T$, if $G \in G^K$;

(II) for every $G \in G$, $\sigma[T|E(G)] \subset \overline{G}$;

(III) for any $\{G_i\}_{i=0}^n \in \text{cov } \sigma(T)$ with $G_0 \in V_\infty$, $\{G_i\}_{i=1}^n \subset G^K$;

$$X = \sum_{i=0}^n E(G_i).$$

It follows from (I) and (II), that $E(\emptyset) = \{0\}$, and (III) implies that, for every open $G \in V_\infty$ with $G \supset \sigma(T)$, $E(G) = X$, in particular, $E(\mathbb{C}) = X$. Evidently, T endowed with a spectral resolvent has the SDP.

11.2. LEMMA. Let T have a spectral resolvent E. For any $G \in G$, the following inclusions hold:

(i) $E(G) \subset X(T,\overline{G})$,
in particular, $E(G) = \Xi(T,\overline{G})$ if $G \in G^K$ and $\partial G \subset \rho(T)$;

(ii) $E(G) \subset \Xi(T,\overline{G})$ if $G \in G^K$;

(iii) $\overline{G \cap \sigma(T)} \subset \sigma[T|E(G)]$

PROOF. (i): For every $x \in E(G)$, we have $\sigma(x,T) \subset \sigma[T|E(G)] \subset \overline{G}$ and hence $x \in X(T,\overline{G})$. Now assume that $G \in G^K$ and $\partial G \subset \rho(T)$. Then ∂G is compact and $\sigma(T)$ being closed, there is a positive number $r = d[\partial G,\sigma(T)]$. Let

$$H = \{\lambda : d(\lambda,\partial G) > \frac{r}{2}\} \cap (\overline{G})^c.$$

Then $\{G,H\} \in \operatorname{cov} \sigma(T)$ and $X = E(G) + E(H)$. It is easily seen that $E(G) \subset \Xi(T,\overline{G})$. To prove the opposite inclusion, let $x \in \Xi(T,\overline{G})$. For any representation

$$x = x_G + x_H \text{ with } x_G \in E(G), \quad x_H \in E(H),$$

we have

$$\sigma(x_H,T) \subset [\sigma(x,T) \cup \sigma(x_G,T)] \cap \overline{H} \subset \overline{G} \cap \overline{H} = \emptyset.$$

Since $x_H = x - x_G \in \Xi(T,\overline{G}) + E(G) = \Xi(T,\overline{G})$, for $\Xi(T,\overline{G}) = Y$, it follows from 4.33 that $x_H = 0$. Therefore, $x = x_G \in E(G)$ and this implies that $\Xi(T,\overline{G}) \subset E(G)$.

(ii): Let $G \in G^K$. Then $E(G) \subset \mathcal{D}_T$ and $T|E(G)$ is bounded. By (i), $x \in E(G)$ implies that $x \in X(T,\overline{G})$ and

$$\lim_{\lambda\to\infty} x(\lambda) = \lim_{\lambda\to\infty} R[\lambda;T|E(G)]x = 0$$

Then $x \in \Xi(T,\overline{G})$, by 5.11 and (ii) follows.

(iii): We may assume that $G \neq \mathbb{C}$. Let $\lambda \in G \cap \sigma(T)$. Choose $H \in G$ such that $\{G,H\} \in \operatorname{cov} \sigma(T)$ and $\lambda \notin \overline{H}$. Then $X = E(G) + E(H)$ and

$$\sigma(T) \subset \sigma[T|E(G)] \cup \sigma[T|E(H)].$$

Since $\lambda \in \sigma[T|E(G)] - \overline{H}$, one has $\lambda \in \sigma[T|E(G)]$ and hence (iii) holds. □

It follows from the foregoing lemma that if T has a spectral resolvent E, then it has a maximal spectral resolvent E_m in the following sense:

$$E(G) \subset E_m(G) = \begin{cases} \Xi(T,\overline{G}), & \text{if } G \in G^K; \\ X(T,\overline{G}), & \text{if } G \text{ is unbounded.} \end{cases} \tag{11.1}$$

Given $F(\neq \emptyset)$, let

$$G_{F,\lambda} = \{\mu : |\mu - \lambda| > \frac{1}{2} d(\lambda,F)\}.$$

For $F(\neq \mathbb{C}) \in F$, define

$$G_F = \begin{cases} \{G_{F,\lambda} : \lambda \in F^c\}, & \text{if } F \neq \emptyset; \\ G \cap V_\infty, & \text{if } F = \emptyset. \end{cases} \tag{11.2}$$

11.3. LEMMA. Let T have a spectral resolvent E. Then, for every $F(\neq \mathbb{C}) \in \mathcal{F}$,

$$X(T,F) = \bigcap \{E(G) : G \in \mathcal{G}_F\}. \tag{11.3}$$

PROOF. First, assume that $\emptyset \neq F \neq \mathbb{C}$. Let $\lambda \in F^c$ and define

$$G = \{\mu : |\mu - \lambda| > \frac{1}{2} d(\lambda,F)\}, \quad H = \{\mu : |\mu - \lambda| < \frac{3}{4} d(\lambda,F)\}.$$

Then $\{G,H\} \in \operatorname{cov} \sigma(T)$ with $G \in V_\infty$, $H \in G^K$ and $\overline{H} \cap F = \emptyset$. We have

$$X = E(G) + E(H), \quad E(H) \subset \mathcal{D}_T. \tag{11.4}$$

Furthermore, $X/E(G)$ and $E(H)/E(G) \cap E(H)$ are topologically isomorphic; $\hat{T} = T/E(G)$ and $\tilde{T} = [T|E(H)]/E(G) \cap E(H)$ are similar. It follows from

$$\sigma(\tilde{T}) \subset \sigma[T|E(H)] \cup \sigma[T|E(G) \cap E(H)]$$

and from

$$\sigma[T|E(G) \cap E(H)] \subset \overline{H}, \tag{11.5}$$

that $\sigma(\hat{T}) = \sigma(\tilde{T}) \subset \overline{H}$. Note that (11.5) is a consequence of

$$\rho[T|E(G) \cap E(H)] \supset \rho_\infty[T|E(H)] \supset (\overline{H})^c.$$

Let $x \in X(T,F)$. For $\lambda \in F^c$, one has $(\lambda - T)x(\lambda) = x$ and hence

$$\hat{x} = \frac{1}{2\pi i} \int_\Gamma R(\lambda;\hat{T})\hat{x}\,d\lambda = \frac{1}{2\pi i} \int_\Gamma R(\lambda;\hat{T})(\lambda-\hat{T})\hat{x}(\lambda)\,d\lambda = \frac{1}{2\pi i} \int_\Gamma \hat{x}(\lambda)\,d\lambda = \hat{0},$$

where $\Gamma = \{\mu : |\mu - \lambda| = \frac{4}{5} d(\lambda,F)\}$. Therefore $x \in E(G)$ and since $G \in \mathcal{G}_F$,

$$X(T,F) \subset \bigcap \{E(G) : G \in \mathcal{G}_F\}. \tag{11.6}$$

To go in the other direction, let $x \in \bigcap \{E(G) : G \in \mathcal{G}_F\}$. For every $G \in \mathcal{G}_F$, we have

$$\sigma(x,T) \subset \sigma[T|E(G)] \subset \overline{G}$$

and hence

$$\sigma(x,T) \subset \bigcap \{\overline{G} : G \in \mathcal{G}_F\} = F.$$

Consequently, $x \in X(T,F)$ and hence the opposite of (11.6) follows.

If $F = \emptyset$, then for every $G \in \mathcal{G} \cap V_\infty$, there is $H \in G^K$ such that $\{G,H\} \in \operatorname{cov} \sigma(T)$. Then, we recapture the decomposition (11.4) and obtain

$$X(T,\emptyset) = \bigcap \{E(G) : G \in \mathcal{G} \cap V_\infty\}. \ \square$$

If we drop condition (I) from 11.1, the resulting weaker concept of *prespectral resolvent* provides new criteria for T to have the SDP.

11.4. DEFINITION. A map $\tilde{E} : G \to \text{Inv } T$ is called a prespectral resolvent of T if it satisfies the following conditions:

(i) for every $G \in G$, $\sigma[T|\tilde{E}(G)] \subset \bar{G}$:

(ii) for any $\{G_i\}_{i=0}^n \in \text{cov } \sigma(T)$ with $G_0 \in V_\infty$, $\{G_i\}_{i=1}^n \subset G^K$,

$$X = \sum_{i=0}^n \tilde{E}(G_i).$$

In contrast to E, we may have $\tilde{E}(\emptyset) \neq \{0\}$, but for every open $G \in V_\infty$ with $G \supset \sigma(T)$, (ii) implies that $\tilde{E}(G) = X$, in particular, $\tilde{E}(\mathbb{C}) = X$. For $T \in B(X)$, the concepts of prespectral and spectral resolvent are identical.

11.5. DEFINITION. Given T, let Z be the set of all $x \in X$ which have the property that, for every $\lambda_0 \in \mathbb{C}$, there is a neighborhood δ of λ_0 and there is an analytic function $f : \delta \to \mathcal{D}_T$ satisfying

$$(\lambda-T)f(\lambda) = x \text{ on } \delta.$$

11.6. THEOREM. Let $\tilde{E}$ be a prespectral resolvent of T. The following assertions are equivalent:

(I) $Z \subset \tilde{E}(G)$, for every $G \in V_\infty$;

(II) there is a spectral resolvent E of T with the properties

$$E(G) \subset \tilde{E}(G) \text{ if } G \in G^K \text{ and } E(G) = \tilde{E}(G) \text{ if } G \text{ is unbounded.} \tag{11.7}$$

PROOF. (I) => (II): Let $H \in G^K$. Then $\sigma[T|\tilde{E}(H)]$ is compact and in view of 3.9, we have

$$\tilde{E}(H) = \tilde{E}'(H) \oplus W, \tag{11.8}$$

where $\tilde{E}'(H)$, $W \in \text{Inv } T$, $\tilde{E}'(H) \subset \mathcal{D}_T$, $\sigma[T|\tilde{E}'(H)] = \sigma[T|\tilde{E}(H)]$, $\sigma(T|W) = \emptyset$. Then $T|\tilde{E}'(H)$ is bounded and $W \subset Z$. If $G \in V_\infty$ then, for each $H \in G^K$, the summand W from (11.8) is contained in $\tilde{E}(G)$. In fact, by hypothesis, $W \subset Z \subset \tilde{E}(G)$. Define $E : G \to \text{Inv } T$, by

$$E(G) = \begin{cases} \tilde{E}'(G), & \text{if } G \in G^K; \\ \tilde{E}(G), & \text{if } G \text{ is unbounded.} \end{cases}$$

Then, for $\{G_i\}_{i=0}^n \in \text{cov } \sigma(T)$ with $G_0 \in V_\infty$ and $\{G_i\}_{i=1}^n \subset G^K$, we have

$$X = \sum_{i=0}^n \tilde{E}(G_i) = \tilde{E}(G_0) + \sum_{i=1}^n \tilde{E}'(G_i) = \sum_{i=0}^n E(G_i);$$

$$\sigma[T|E(G_0)] = \sigma[T|\tilde{E}(G_0)] \subset \overline{G}_0;$$

$$\sigma[T|E(G_i)] = \sigma[T|\tilde{E}'(G_i)] = \sigma[T|\tilde{E}(G_i)] \subset \overline{G}_i, \ 1 \leq i \leq n.$$

Thus, E is a spectral resolvent of T that satisfies conditions (11.7).

(II) => (I): Let E be any spectral resolvent of T that satisfies conditions (11.7). By 11.3, $X(T,\emptyset) = \bigcap \{E(G) : G \in V_\infty\}$. Since, for every $G \in V_\infty$, $E(G) = \tilde{E}(G)$, we have $Z = X(T,\emptyset) \subset \tilde{E}(G)$ if $G \in V_\infty$. □

The following properties are immediate consequences of the foregoing theorem.

11.7. COROLLARY. If $Z = \{0\}$, then every prespectral resolvent of T is a spectral resolvent of T.

11.8. COROLLARY. If T has a prespectral resolvent $\tilde{E}$ that satisfies condition (I) of 11.6, then T has the SDP.

The notion of *support* of $x \in X$, related to a specific set-function (spectral capacity), has its roots in a work of Apostol [Ap.1968, c] and has been carried over to spectral resolvents by Shulberg [Sh.1979]. In the following few passages we expand a little in this circle of ideas.

11.9. DEFINITION. Let T have a spectral resolvent E. For each $x \in X$, define the support of x with respect to E, by

$$\text{Supp}_E(x) = \bigcap_{x \in E(G)} \overline{G}.$$

Furthermore, define the support of E, by

$$\text{Supp } E = \bigcap_{E(G)=X} \overline{G}$$

11.10. THEOREM. Let T have a spectral resolvent E. Then,

(11.9) $\quad \text{Supp}_E(x) = \sigma(x,T)$ for all $x \in X$;

(11.10) $\quad \text{Supp } E = \sigma(T)$.

PROOF. Since $x \in E(G)$ implies that $\sigma(x,T) \subset \overline{G}$, we have $\sigma(x,T) \subset \text{Supp}_E(x)$. Conversely, denote $F = \sigma(x,T)$. Then $x \in X(T,F)$ and it follows from (11.3) that $x \in \bigcap_{G \in G_F} E(G)$. Therefore,

$$\text{Supp}_E(x) \subset \bigcap_{G \in G_F} \overline{G} = F = \sigma(x,T).$$

Next, we prove (11.10). For every $x \in X$, we have $x \in \bigcap_{E(G)=X} E(G)$ and hence

$$\sigma(x,T) \subset \bigcap_{E(G)=X} \overline{G} = \text{Supp } E.$$

Then, by 2.7, we obtain

$$\sigma(T) = \bigcup_{x \in X} \sigma(x,T) \subset \text{Supp } E.$$

Conversely, put $F = \sigma(T)$, use (11.2) and obtain

$$X = X(T,F) = \bigcap_{G \in G_F} E(G).$$

Since, for every $G \in G_F$, we have $E(G) = X$, it follows that

$$\text{Supp } E = \bigcap_{E(G)=X} \overline{G} \subset \bigcap_{G \in G_F} \overline{G} = F = \sigma(T). \quad \square$$

A slightly modified version of the spectral resolvent concept produces a surprising result. Shulberg [Sh.1979] introduced the concept of *open spectral resolvent*, by substituting condition $\sigma[T|E(G)] \subset \overline{G}$ by the more restrictive $\sigma[T|E(G)] \subset G$ and showed that $T \in B(X)$ has an open spectral resolvent iff $\sigma(T)$ is finite [ibid., Proposition 13]. What does this property become if we drop the boundedness assumption?

11.11. DEFINITION. Given T, a map $E : G \to \text{Inv } T$ is said to be an open spectral resolvent of T if the following conditions are satisfied.

(i) $E(G) \subset \mathcal{D}_T$, if $G \in G^K$;

(ii) for every $G \in G$, $\sigma[T|E(G)] \subset G$;

(iii) for any $\{G_i\}_{i=0}^n \in \text{cov } \sigma(T)$ with $G_0 \in V_\infty$, $\{G_i\}_{i=1}^n \subset G^K$;

$$X = \sum_{i=0}^n E(G_i).$$

Evidently, if T is endowed with an open spectral resolvent then T has the SDP.

11.12. THEOREM. T has an open spectral resolvent iff $\sigma(T)$ consists of isolated points.

PROOF. (Only if): Suppose that T has an open spectral resolvent E and $\sigma(T)$ has a cluster point λ_0. There exists a sequence $\{\lambda_n\} \subset \sigma(T)$ such that $\lambda_n \to \lambda_0$, $\lambda_n \neq \lambda_0$. Let $G \in G^K$ be such that $\{\lambda_n\} \subset G$ and $\lambda_0 \in \partial G$.

Condition $\sigma[T|E(G)] \subset G$ implies that $\lambda_0 \in \rho[T|E(G)]$ and hence $\lambda_m \in \rho[T|E(G)]$ for m sufficiently large. Let $G_0 \in V_\infty$ be such that $\sigma[T|E(G)] \cap G_0 = \emptyset$, $\lambda_m \notin G_0$ and $\{G_0, G\} \in \text{cov}\, \sigma(T)$. Then

$$(11.11) \qquad \lambda_m \in \sigma(T) \cap \rho[T|E(G)] \cap \rho[T|E(G_0)].$$

In view of $X = E(G_0) + E(G)$, we have

$$\sigma(T) \subset \sigma[T|E(G_0)] \cup \sigma[T|E(G)].$$

Thus λ_m is an element of, at least, one of the sets $\sigma[T|E(G_0)]$, $\sigma[T|E(G)]$; but this contradicts (11.11).

(If): Assume that $\sigma(T)$ consists of isolated points. Then, by the functional calculus, T has the SDP. For any $G \in \mathcal{G}$, one of the following cases is possible:

(1) $G \in \mathcal{G}^K$, then $G \cap \sigma(T)$ is finite;

(2) G is unbounded, then $G \cap \sigma(T)$ is, at most infinite countable.

In case (2), let $\{\lambda_n\} \subset G \cap \sigma(T)$. Then $\{\lambda_n\}$ has no finite cluster point and hence $\lambda_n \to \infty$. In either case, $G \cap \sigma(T)$ is closed. Therefore, $X(T,G) = X[T, G \cap \sigma(T)]$ is closed. Define

$$(11.12) \qquad E(G) = \begin{cases} \Xi(T,G), & \text{if } G \in \mathcal{G}^K; \\ X(T,G), & \text{if } G \text{ is unbounded.} \end{cases}$$

Then E satisfies conditions (i) and (iii) of 11.11. Moreover,

$$\sigma[T|\Xi(T,G)] = \sigma\{T|\Xi[T, G \cap \sigma(T)]\} \subset G \cap \sigma(T) \subset G, \text{ if } G \in \mathcal{G}^K;$$

$$\sigma[T|X(T,G)] = \sigma\{T|X[T, G \cap \sigma(T)]\} \subset G \cap \sigma(T) \subset G, \text{ if } G \text{ is un-}$$

bounded. Thus, in both cases, we have $\sigma[T|E(G)] \subset G$ and this is condition (ii) of 11.11. □

§.12. MONOTONIC SPECTRAL RESOLVENTS.

Given T, for the maximal spectral resolvent (11.1) E_m of T, $\overline{G}_1 \subset G_2$ implies $E_m(G_1) \subset E_m(G_2)$. This property may not hold for any spectral resolvent E, but if it does then E gives rise to some interesting implications.

12.1. DEFINITION. Given $T \in B(X)$, a spectral resolvent E of T is said to be *monotonic* if $G_1, G_2 \in \mathcal{G}$ and $\overline{G}_1 \subset G_2$ imply $E(G_1) \subset E(G_2)$.

12.2. DEFINITION. Given $T \in B(X)$, a spectral resolvent E of T is said to be *strongly monotonic* if $G, G_1, G_2 \in \mathcal{G}$ and $\overline{G}_1 \cap \overline{G}_2 \subset G$ imply $E(G_1) \cap E(G_2) \subset E(G)$.

Evidently, every strongly monotonic spectral resolvent is monotonic. If $T \in B(X)$ has the SDP then E_m is strongly monotonic.

Since clearly, $\overline{X(T,G)} \subset X(T,\overline{G})$ where the inclusion may be proper, some spectral resolvents E may be such that

(12.1) $$\overline{X(T,G)} \subset E(G) \subset X(T,\overline{G}) \quad \text{for all } G \in \mathcal{G}.$$

12.3. THEOREM. Let $T \in B(X)$ have a spectral resolvent E. Then E is strongly monotonic iff (12.1) holds.

PROOF. (If): Let $G, G_1, G_2 \in \mathcal{G}$ with $\overline{G}_1 \cap \overline{G}_2 \subset G$. Then, with the help of (12.1), we obtain

$$E(G_1) \cap E(G_2) \subset X(T,\overline{G}_1) \cap X(T,\overline{G}_2) = X(T,\overline{G}_1 \cap \overline{G}_2) \subset \overline{X(T,G)} \subset E(G).$$

(Only if): We run this section of the proof in two parts.

Part A. Given $G \in \mathcal{G}$, let $x \in X(T,G)$. Choose $\{G_1, G_2\} \in \text{cov}\,\sigma(T)$ such that $\sigma(x,T) \subset G_1 \subset \overline{G}_1 \subset G$ and $\sigma(x,T) \cap \overline{G}_2 = \emptyset$. There is a representation of x,

(12.2) $$x = x_1 + x_2, \quad x_i \in E(G_i), \quad i=1,2.$$

The local spectra of x_1 and x_2 are contained in some pertinent sets

$$\sigma(x_1,T) \subset \sigma(x,T) \cup (\overline{G}_1 \cap \overline{G}_2), \quad \sigma(x_2,T) \subset \overline{G}_1 \cap \overline{G}_2.$$

For $\lambda \in \rho(x,T) \cap (\overline{G}_1 \cap \overline{G}_2)^c = H$, we have

(12.3) $$x(\lambda) = x_1(\lambda) + x_2(\lambda).$$

For a bounded Cauchy neighborhood Δ of $\sigma(x,T)$ with $\overline{\Delta} \cap \overline{G}_2 = \emptyset$, we have

(12.4) $$x = \frac{1}{2\pi i} \int_{\partial\Delta} x(\lambda)d\lambda = \frac{1}{2\pi i} \int_{\partial\Delta} x_1(\lambda)d\lambda.$$

By 3.11, for every $\lambda_0 \in \partial\Delta$, there exists a neighborhood $V \subset H$ of λ_0 and there are functions $f_i : V \to E(G_i)$, $i=1,2$ analytic on V such that

(12.5) $$x_1(\lambda) = f_1(\lambda) + f_2(\lambda) \quad \text{on } V.$$

It follows from

$$(\lambda - T)x_1(\lambda) = x_1 \quad \text{on } \rho(x_1,T) \supset H \supset V,$$

that $g : V \to E(G_1) \cap E(G_2)$, defined by

$$g(\lambda) = x_1 - (\lambda-T)f_1(\lambda) = (\lambda-T)f_2(\lambda)$$

is analytic on V.

Part B. Let $K \in G$ be such that

$$\overline{G}_1 \cap \overline{G}_2 \subset K \subset \overline{K} \subset G, \quad \overline{K} \cap \sigma(x,T) = \emptyset \quad \text{and} \quad V \cap \overline{K} = \emptyset.$$

E being strongly monotonic, we have $g(\lambda) \in E(K)$ on V. The function $h : V \to E(K)$, defined by $h(\lambda) = R[\lambda;T|E(K)]g(\lambda)$, is analytic on V and satisfies property

$$(\lambda-T)h(\lambda) = (\lambda-T)f_2(\lambda) \quad \text{on } V.$$

By the SVEP, $f_2(\lambda) = h(\lambda) \in E(K)$ on V. E being monotonic, we have

$$x_1(\lambda) \in E(G_1) + E(K) \subset E(G) \quad \text{on } V.$$

In particular, $x_1(\lambda_0) \in E(G)$. Since λ_0 is arbitrary on $\partial\Delta$, (12.4) implies that $x \in E(G)$. Consequently, inclusion $\overline{X(T,G)} \subset E(G)$ follows. □

12.4. THEOREM. Let $T \in B(X)$ have a spectral resolvent E. E is strongly monotonic iff conditions: $G_1, G_2 \in G$, $\overline{G}_1 \subset G_2$ and $x \in E(G_1)$ imply

$$\{x(\lambda) : \lambda \in \rho(x,T)\} \subset E(G_2).$$

PROOF. (Only if): Given $G_1, G_2 \in G$ with $\overline{G}_1 \subset G_2$. E being strongly monotonic, 12.3 implies

$$E(G_1) \subset X(T,\overline{G}_1) \subset X(T,G_2) \subset E(G_2). \tag{12.6}$$

Let $x \in E(G_1)$ be given. Then $x \in X(T,\overline{G}_1)$ and the latter being a spectral maximal space of T,(12.6) implies

$$\{x(\lambda) : \lambda \in \rho(x,T)\} \subset X(T,\overline{G}_1) \subset E(G_2).$$

(If): As in Part A of the proof of 12.3, for $G \in G$, $x \in X(T,G)$, choose $\{G_1,G_2\} \in \operatorname{cov} \sigma(T)$ with

$$\sigma(x,T) \subset G_1 \subset \overline{G}_1 \subset G, \quad \sigma(x,T) \cap \overline{G}_2 = \emptyset,$$

and a Cauchy neighborhood Δ of $\sigma(x,T)$ with $\overline{\Delta} \subset (\overline{G}_1 \cap \overline{G}_2)^c$, to obtain

$$x = \frac{1}{2\pi i}\int_{\partial\Delta} x(\lambda)d\lambda = \frac{1}{2\pi i}\int_{\partial\Delta} x_1(\lambda)d\lambda. \tag{12.7}$$

For x_1 and $x_1(\lambda)$, we refer to representations (12.2) and (12.3), respectively. Since $x_1 \in E(G_1)$ and $\overline{G}_1 \subset G$, the hypothesis implies

$$\{x_1(\lambda) : \lambda \in \rho(x,T)\} \subset E(G).$$

Then $x \in E(G)$, by (12.7). Thus $\overline{X(T,G)} \subset E(G)$ and hence E is strongly monotonic, by 12.3. □

A further characterization of a strongly monotonic spectral resolvent can be obtained in terms of a localization property of the spectral resolvent. To this end, we generalize the concept of *almost localized spectrum.*

12.5. DEFINITION. A spectral resolvent E is said to be *almost localized* if $G, G_1, G_2 \in \mathcal{G}$ and $\overline{G} \subset G_1 \cup G_2$ imply $E(G) \subset E(G_1) + E(G_2)$.

By induction, the defining property of an almost localized spectral resolvent can be extended to any finite system $\{G, G_1, \ldots, G_n\} \subset \mathcal{G}$ in the sense that

$$\overline{G} \subset \bigcup_{i=1}^{n} G_i \quad \text{implies} \quad E(G) \subset \sum_{i=1}^{n} E(G_i). \tag{12.8}$$

It is easily seen that if E is almost localized then it is monotonic.

12.6. THEOREM. Let $T \in B(X)$ have a spectral resolvent E. Then E is strongly monotonic iff E is almost localized.

PROOF. In view of 12.3, it suffices to show that the following conditions

(I) $\overline{X(T,G)} \subset E(G)$ for all $G \in \mathcal{G}$;

(II) $G, G_1, G_2 \in \mathcal{G}$ and $\overline{G} \subset G_1 \cup G_2$ imply $E(G) \subset E(G_1) + E(G_2)$,

are equivalent.

(I) => (II): Let G, G_1 and G_2 be as asserted in (II). There are open sets H_1, H_2 with $\overline{H}_1 \subset G_1$, $\overline{H}_2 \subset G_2$ and $\overline{G} \subset H_1 \cup H_2$. Then, by quoting 6.2, one obtains successively

$$E(G) \subset X(T,\overline{G}) \subset X(T,\overline{H}_1) + X(T,\overline{H}_2) \subset \overline{X(T,G_1)} + \overline{X(T,G_2)} \subset E(G_1) + E(G_2).$$

(II) => (I): Given $G \in \mathcal{G}$, let $x \in X(T,G)$. Let H_0 be a relatively compact open neighborhood of $\sigma(T)$. Then $x \in X = E(H_0)$ and $\sigma(x,T) \subset \sigma(T) \subset H_0$. Further, let ε be arbitrary such that

$$0 < \varepsilon < \sup_{\lambda \in \partial H_0} d[\lambda, \sigma(x,T)]$$

and define the open sets

$$H = \{\lambda : d[\lambda, \sigma(x,T)] < \varepsilon\},$$

$$H' = \{\lambda : d(\lambda, H_0) < \tfrac{\varepsilon}{6}\}.$$

For every $\lambda \in \overline{H}' \cap H^c$, let $D_\lambda = \{\mu : |\mu - \lambda| < \frac{\varepsilon}{3}\}$. Then $\{D_\lambda : \lambda \in \overline{H}' \cap H^c\}$ is an open cover of $\overline{H}' \cap H^c$. Since $\overline{H}' \cap H^c$ is compact, there is a finite collection $\{\lambda_1, \lambda_2, \ldots, \lambda_n\} \subset \overline{H}' \cap H^c$ such that

$$\overline{H}' \cap H^c \subset \bigcup_{i=1}^{n} D_i, \quad \text{where } D_i = D_\lambda \text{ for } \lambda = \lambda_i.$$

For $1 \leq i \leq n$, define

$$K_i = \{\mu : |\mu - \lambda_i| < \tfrac{2\varepsilon}{3}\}, \quad \Delta_i = \{\mu : |\mu - \lambda_i| < \tfrac{\varepsilon}{2}\}.$$

Clearly, $\overline{K}_i \cap \sigma(x,T) = \emptyset$, $1 \leq i \leq n$. Put

$$H_1 = \{\lambda : d(\lambda, H_0) < \tfrac{\varepsilon}{9n}\} - \overline{\Delta}_1$$

and note that $\overline{H}_1 \cap \overline{D}_1 = \emptyset$. Since $\overline{H}_0 \subset H_1 \cup \overline{\Delta}_1 \subset H_1 \cup K_1$, we have

$$x \in E(H_0) \subset E(H_1) + E(K_1).$$

To save repetition, for $G_1 = H_1$ and $G_2 = K_1$, follow Part A from the proof of 12.3. Note that the boundary $\partial\Delta$ of the Cauchy domain Δ in Part A, verifies inclusions

$$\partial\Delta \subset \rho_\infty[T|E(K_1)] \subset \rho[T|E(H_1) \cap E(K_1)].$$

The function $h : V \to E(H_1) \cap E(K_1)$, defined by

$$h(\lambda) = R[\lambda; T|E(H_1) \cap E(K_1)]g(\lambda),$$

is analytic on V and verifies

$$(\lambda - T)h(\lambda) = (\lambda - T)f_2(\lambda) \quad \text{on } V.$$

This implies

$$f_2(\lambda) = h(\lambda) \in E(H_1) \cap E(K_1) \quad \text{on } V.$$

Thus, with reference to Part A, (12.5) implies that $x_1(\lambda) \in E(H_1)$ on V and hence $x_1(\lambda_0) \in E(H_1)$. Since $\lambda_0 \in \partial\Delta$ is arbitrary, $x \in E(H_1)$, by (12.4).

Inductively define

$$H_k = \{\lambda : d(\lambda, H_{k-1}) < \tfrac{\varepsilon}{9n}\} - \overline{\Delta}_k, \quad 1 \leq k \leq n.$$

Then $\{H_k, K_k\}$ covers $\overline{H}_{k-1}$ and $\overline{H}_k \cap \overline{D}_i = \emptyset$, $1 \leq i \leq k$. In view of (II),

$$E(H_{k-1}) \subset E(H_k) + E(K_k),$$

and the inductive hypothesis $x \in E(H_{k-1})$ gives

$$x \in E(H_k) + E(K_k).$$

As in case $k=1$, use Part A and, for a conveniently defined function $h: V \to E(H_k) \cap E(K_k)$, obtain $x \in E(H_k)$. Thus, by the inductive process, one obtains an open set H_n, with the following properties:

$$x \in E(H_n) \text{ and } \overline{H}_n \subset H' - (\bigcup_{i=1}^{n} \overline{D}_i) \subset H.$$

E being monotonic, $E(H_n) \subset E(H)$ and hence $x \in E(H)$. Since ε is arbitrarily small, we may choose it such that $\overline{H} \subset G$. Then $E(H) \subset E(G)$ and hence $x \in E(G)$. Since $x \in X(T,G)$ is arbitrary, we obtain

$$\overline{X(T,G)} \subset E(G). \quad \square$$

§.13. ANALYTICALLY INVARIANT SPECTRAL RESOLVENTS.

Analytic vector valued functions play a major role in the spectral theory, one important outgrowth being the analytically invariant subspace concept. Some of the properties of analytically invariant subspaces were discussed in §.4 and subsequently applied to the spectral duality theory (§.10). Many of the hidden implications of analytic invariance come to the surface in the framework of *analytically invariant spectral resolvents*.

A few preliminary properties of analytically invariant subspaces will open the path to the present study.

13.1. THEOREM. Given T with the SDP, let Y be a subspace of X. The following assertions are equivalent:

(I) Y is a μ-space of T and, for every compact $F \subset \operatorname{Int} \sigma(T|Y)$, $\Xi(T,F) \subset Y$;

(II) $Y \in \operatorname{Inv} T$, $\hat{T} = T/Y$ is closed, and

(13.1) $$\sigma(\hat{T}) \cap \operatorname{Int} \sigma(T|Y) = \emptyset.$$

Furthermore, if either of conditions (I) or (II) holds, then Y is analytically invariant under T.

PROOF. (I) => (II): If $\operatorname{Int} \sigma(T|Y) = \emptyset$, then $\hat{T}$ is closed by 5.16. Consequently, the properties expressed by (II) hold. Suppose that $\operatorname{Int} \sigma(T|Y) \neq \emptyset$. Let $\lambda \in \operatorname{Int} \sigma(T|Y)$ and let $G \in G^K$ be such that $\lambda \in G \subset \overline{G} \subset \operatorname{Int} \sigma(T|Y)$. Choose $G_0 \in V_\infty$ such that $\{G_0, G\} \in \operatorname{cov} \sigma(T)$ and $\lambda \notin \overline{G}_0$. Then

$$X = X(T,\overline{G}_0) + \Xi(T,\overline{G}) = X(T,\overline{G}_0) + Y.$$

Since Y is a μ-space of T, use 4.9 to write

$$\sigma[T|Y \cap X(T,\overline{G}_0)] = \sigma[(T|Y)|Y(T|Y,\overline{G}_0)] \subset \overline{G}_0,$$

where $Y(T|Y,\cdot)$ denotes the spectral manifold of $T|Y$ in Y. $\overline{G}$ being compact, $\Xi(T,\overline{G}) \subset \mathcal{D}_T$. For $X_0 = X(T,\overline{G}_0)$, $X_1 = \Xi(T,\overline{G})$ and the given Y, 3.2 implies that $\hat{T}$ is closed and $\sigma(\hat{T}) = \sigma(\tilde{T})$, where $\tilde{T} = [T|X(T,\overline{G}_0)]/Y \cap X(T,\overline{G}_0)$. Now $\lambda \notin \overline{G}_0$ implies that $\lambda \notin \sigma[T|X(T,\overline{G}_0)] \cup \sigma[T|Y \cap X(T,\overline{G}_0)]$ and hence $\lambda \in \rho(\tilde{T})$, by 3.1. Since λ is arbitrary in Int $\sigma(T|Y)$, (13.1) holds.

(II) => (I): First, we show that Y is a μ-space of T. Let $x \in Y$. For $\lambda \in \rho(x,T) \cap \text{Int}\,\sigma(T|Y)$, (13.1) implies that $\lambda \in \rho(\hat{T})$ and it follows from $(\lambda-T)x(\lambda) = x$ that $(\lambda-\hat{T})\hat{x}(\lambda) = \hat{0}$. Consequently, $\hat{x}(\lambda) = \hat{0}$ and hence $x(\lambda) \in Y$ on $\rho(x,T) \cap \text{Int}\,\sigma(T|Y)$. On the other hand, $\lambda \in \rho(T|Y)$ implies that $x(\lambda) = R(\lambda;T|Y)x \in Y$. Thus, it follows that $x(\lambda) \in Y$ on $\rho(x,T) - \partial\rho(T|Y)$ and hence $x(\lambda) \in Y$ on $\rho(x,T)$. Hence Y is a μ-space of T, by 4.8.

Next, let $F \subset \text{Int}\,\sigma(T|Y)$ be compact and let $x \in \Xi(T,F)$. To prove the inclusion $\Xi(T,F) \subset Y$, put

$$\hat{f}(\lambda) = \begin{cases} \hat{x}(\lambda) & \text{if } \lambda \in \rho(x,T); \\ R(\lambda;\hat{T})\hat{x}, & \text{if } \lambda \in \rho(\hat{T}). \end{cases}$$

For $\lambda \in \rho(x,T) \cap \rho(\hat{T})$, we have

$$(\lambda-\hat{T})[R(\lambda;\hat{T})\hat{x} - \hat{x}(\lambda)] = \hat{x} - \hat{x} = \hat{0}$$

and since $\lambda-\hat{T}$ is injective, we have $R(\lambda;\hat{T})\hat{x} = \hat{x}(\lambda)$. Therefore, $\hat{f}$ is well-defined. Furthermore, $\sigma(x,T) \subset F \subset \text{Int}\,\sigma(T|Y)$ and (13.1) imply that

$$\sigma(x,T) \cap \sigma(\hat{T}) = \emptyset.$$

Thus, $\rho(x,T) \cup \rho(\hat{T}) = \mathbb{C}$ and hence $\hat{f}$ is entire. Since $\sigma[T|\Xi(T,F)]$ is compact, $T|\Xi(T,F)$ is bounded and

$$\Gamma = \{\lambda : |\lambda| = \|T|\Xi(T,F)\| + 1\} \subset \rho[T|\Xi(T,F)] \subset \rho(x,T).$$

It follows from

$$x = \frac{1}{2\pi i}\int_\Gamma R[\lambda;T|\Xi(T,F)]x\,d\lambda = \frac{1}{2\pi i}\int_\Gamma x(\lambda)\,d\lambda$$

that

$$\hat{x} = \frac{1}{2\pi i}\int_\Gamma \hat{x}(\lambda)\,d\lambda = \frac{1}{2\pi i}\int_\Gamma \hat{f}(\lambda)\,d\lambda = \hat{0}.$$

Consequently, $x \in Y$ and hence $\Xi(T,F) \subset Y$.

Finally suppose that conditions (I) and (II) hold. Put $K = \sigma(T|Y)$ and recall that $Y \subset X(T,K)$. First, we show that $\hat{X}_K = X(T,K)/Y$ is a spectral maximal space of $\hat{T} = T/Y$. Let $\hat{Z} \in \text{Inv}\,\hat{T}$ be such that $\sigma(\hat{T}|\hat{Z}) \subset \sigma(\hat{T}|\hat{X}_K)$. Note that $Z = \{x \in X : x \in \hat{x},\ \hat{x} \in \hat{Z}\} \in \text{Inv}\,T$, use 3.1 and obtain

$$\text{(13.2)} \qquad \sigma(T|Z) \subset \sigma(\hat{T}|\hat{Z}) \cup \sigma(T|Y) \subset \sigma(\hat{T}|\hat{X}_K) \cup \sigma(T|Y).$$

Since $\sigma[T|X(T,K)] \subset K = \sigma(T|Y)$, it follows that

$$\text{(13.3)} \qquad \sigma(\hat{T}|\hat{X}_K) \subset \sigma[T|X(T,K)] \cup \sigma(T|Y) = \sigma(T|Y).$$

Consequently, (13.2) implies $\sigma(T|Z) \subset \sigma(T|Y) = K$ and hence $Z \subset X(T,K)$ or, equivalently, $\hat{Z} \subset \hat{X}_K$. Therefore, $\hat{X}_K$ is a spectral maximal space of $\hat{T}$. In particular, $\hat{X}_K$ is a ν-space of $\hat{T}$ and hence $\sigma(\hat{T}|\hat{X}_K) \subset \sigma(\hat{T})$. Thus

$$\text{(13.4)} \qquad \sigma(\hat{T}|\hat{X}_K) \cap \text{Int}\,\sigma(T|Y) = \emptyset.$$

It follows from (13.3) and (13.4) that

$$\sigma(\hat{T}|\hat{X}_K) \subset \partial K.$$

To see that Y is analytically invariant under T, let $f : \omega_f \to \mathcal{D}_T$ be analytic and satisfy condition

$$\text{(13.5)} \qquad (\lambda - T)f(\lambda) \in Y \quad \text{on an open } \omega_f.$$

We assume that ω_f is connected. Since $Y \subset X(T,K)$ and $X(T,K)$ is analytically invariant under T (Corollary 5.20), (13.5) implies that $f(\lambda) \in X(T,K)$ on ω_f. Also (13.5) gives rise to

$$(\lambda - \hat{T})\hat{f}(\lambda) = \hat{0} \quad \text{on } \omega_f$$

and since $\hat{f}(\lambda) \in \hat{X}_K$, we have $\hat{f}(\lambda) = \hat{0}$ on $\omega_f - \partial K$. Thus $\hat{f}(\lambda) = \hat{0}$ on ω_f, by analytic continuation and hence $f(\lambda) \in Y$ on ω_f. □

The following property is an immediate consequence of the foregoing theorem.

13.2. COROLLARY. Given T with the SDP, let $G \in \mathcal{G}$. If Y is a μ-space of T such that

$$\overline{\Xi(T,G)} \subset Y \subset X(T,\overline{G})$$

then Y is analytically invariant under T.

13.3. COROLLARY. Let T have the SDP. For every $G \in G^K$, $\overline{\Xi(T,G)}$ is analytically invariant under T.

PROOF. It follows from 9.8 (II) and 9.9 that, for $G \in G^K$,

$$\overline{\Xi(T,G)} \in \text{Inv } T, \quad \sigma[T|\overline{\Xi(T,G)}] \subset \overline{G} \quad \text{and} \quad T/\overline{\Xi(T,G)} \quad \text{is closed.}$$

Let $G_0 \in V_\infty$ be an arbitrary open set that contains G^c. Then $\{G_0,G\} \in \text{cov } \sigma(T)$ and the following decomposition holds

$$X = X(T,\overline{G}_0) + \overline{\Xi(T,G)}.$$

It is easily seen that $\sigma[T|X(T,\overline{G}_0) \cap \overline{\Xi(T,G)}] \subset \overline{G}_0$ and hence

$$\sigma[T/\overline{\Xi(T,G)}] \subset \sigma[T|X(T,\overline{G}_0)] \cup \sigma[T|X(T,\overline{G}_0) \cap \overline{\Xi(T,G)}] \subset \overline{G}_0.$$

Since $G_0(\in V_\infty) \supset G^c$ is arbitrary, it follows that

$$\sigma[T/\overline{\Xi(T,G)}] \subsetneq G^c$$

Thus $\overline{\Xi(T,G)}$ is analytically invariant under T, by 13.1 (II). □

13.4. COROLLARY. Let T have the SDP. For every $G \in G^K$, $\overline{X(T,G)}$ is analytically invariant under T.

PROOF. First, we show that $\overline{X(T,G)}$ is invariant under T. Let $x \in \overline{X(T,G)}$. There is a sequence $\{x_n\} \subset X(T,G)$ converging to x. Since

$$\overline{X(T,G)} \subset X(T,\overline{G}) = \Xi(T,\overline{G}) \oplus X(T,\emptyset), \tag{13.6}$$

we have the following representations

$$x = x_0 + x_1 \quad \text{with} \quad x_0 \in \Xi(T,\overline{G}), \quad x_1 \in X(T,\emptyset);$$

$$x_n = x_{n,0} + x_{n,1} \quad \text{with} \quad x_{n,0} \in \Xi(T,\overline{G}), \quad x_{n,1} \in X(T,\emptyset), \quad n \in \mathbb{N}.$$

Since $\sigma(x_{n,0},T) \subset \sigma(x_n,T) \cup \sigma(x_{n,1},T) \subset G$ there is a compact $F_n \subset G$ such that $\sigma(x_{n,0},T) \subset F_n$. It follows from 5.13 and 9.7 that

$$x_{n,0} \in \Xi(T,\overline{G}) \cap X(T,F_n) = \Xi(T,\overline{G} \cap F_n) \subset \Xi(T,G).$$

By the direct sum (13.6), $x_{n,0} \to x_0$ and hence $x_0 \in \overline{\Xi(T,G)}$. Consequently, $\overline{X(T,G)} \subset \overline{\Xi(T,G)} \oplus X(T,\emptyset)$, and the opposite inclusion being evident, we obtain

$$\overline{X(T,G)} = \overline{\Xi(T,G)} \oplus X(T,\emptyset) \tag{13.7}$$

In view of 9.8 (II) and 9.9, $\overline{X(T,G)} \in \text{Inv } T$.

Next, we show that, for $x \in \overline{X(T,G)}$ and each $\lambda \in \rho(x,T)$, we have $x(\lambda) \in \overline{X(T,G)}$. With the help of (13.7), write the representation of $x \in \overline{X(T,G)}$, as follows

$$x = x_0 + x_1 \quad \text{with} \quad x_0 \in \overline{\Xi(T,G)}, \quad x_1 \in X(T,\emptyset).$$

Denote by $x(\cdot)$, $x_0(\cdot)$, $x_1(\cdot)$ the local resolvents of x, x_0, x_1, respectively. Since $x_1(\cdot)$ is an entire function, $\sigma(x,T) = \sigma(x_0,T)$. We have

$$(13.8) \qquad x(\lambda) = x_0(\lambda) + x_1(\lambda) \quad \text{for} \quad \lambda \in \rho(x,T).$$

It follows from $\sigma[x_1(\lambda),T] = \sigma(x_1,T) = \emptyset$ that $x_1(\lambda) \in X(T,\emptyset)$ for all $\lambda \in \mathbb{C}$. Furthermore, for $\lambda \in \rho(x_0,T)$, 13.3 and 4.11 imply that $x_0(\lambda) \in \overline{\Xi(T,G)}$. Then, for every $\lambda \in \rho(x,T)$, $x(\lambda) \in \overline{X(T,G)}$, by (13.8) and (13.7). Therefore $\overline{X(T,G)}$ is a μ-space of T, by 4.8, and hence $\overline{X(T,G)}$ is analytically invariant under T, by 13.2. □

13.5. COROLLARY. Let T be densely defined and have the SDP. For every $G \in V_\infty$, $\overline{X(T,G)}$ is analytically invariant under T.

PROOF. It follows from 9.8 (I), that $T/\overline{X(T,G)}$ is bounded and $X(T,G)^a = \Xi^*(T^*,G^c)$. The remainder of the proof is analogous to that of 13.3. Indeed, we have

$$\sigma[T/\overline{X(T,G)}] = \sigma[T/{}^a\Xi^*(T^*,G^c)] = \sigma[T^*|\Xi^*(T^*,G^c)] \subset G^c.$$

On the other hand, $\sigma[T|\overline{X(T,G)}] \subset \sigma[T|X(T,\overline{G})] \subset \overline{G}$ implies that

$$\text{Int}\, \sigma[T|\overline{X(T,G)}] \subset G$$

and hence $Y = \overline{X(T,G)}$ satisfies conditions (II) of 13.1. Thus $\overline{X(T,G)}$ is analytically invariant under T. □

It is expedient to cast the analytic invariance in terms of spectral resolvents.

13.6. DEFINITION. Given T, a spectral resolvent E of T is said to be analytically invariant under T if, for every $G \in \mathcal{G}$, $E(G)$ is analytically invariant under T.

For notational convenience, we shall consider two distinct classes of open sets:

$G \in \mathcal{G}_\alpha \subset \mathcal{G}$ if either G is relatively compact or $G \in V_\infty$;

$G \in \mathcal{G}_\beta \subset \mathcal{G}$ if $G \notin \mathcal{G}_\alpha$, i.e. if G is unbounded but $G \notin V_\infty$.

13.7. THEOREM. Given T with the spectral resolvent E, let $G \in \mathcal{G}_\beta$. The following assertions are equivalent:

(I) $E(G)$ is analytically invariant under T;

(II) $E(G)$ is a μ-space of T;

(III) for each $H \in G$,

(13.9) $$\sigma[T|E(G) \cap E(H)] \subset \overline{H}$$

and, for every compact $F \subset G$,

(13.10) $$\Xi(T,F) \subset E(G);$$

(IV) the coinduced $\hat{T} = T/E(G)$ is closed and $\sigma(\hat{T}) \subset G^c$.

PROOF. (I) => (II): evident.

(II) => (III): First, we show that, for $H \in G$, $E(G) \cap E(H)$ is a ν-space of $T|E(H)$. Let $x \in E(G) \cap E(H)$ and $\lambda \in \rho[T|E(H)]$. Then $x(\lambda) = R[\lambda;T|E(H)]x$ and, since $E(G)$ is a μ-space of T, we have

(13.11) $$R[\lambda;T|E(H)]x = x(\lambda) \in E(G)$$

On the other hand, $R[\lambda;T|E(H)]x \in E(H)$ and hence (13.11) implies

$$R[\lambda;T|E(H)]x \in E(G) \cap E(H),\ \lambda \in \rho[T|E(H)].$$

By 4.2, $E(G) \cap E(H)$ is a ν-space of $T|E(H)$ and hence

$$\sigma[T|E(G) \cap E(H)] \subset \sigma[T|E(H)] \subset \overline{H}.$$

To prove (13.10), let $F \subset G$ be compact and let $x \in \Xi(T,F)$ be arbitrary. Choose $G_0 \in V_\infty$ such that $\{G_0,G\} \in \text{cov}\, \sigma(T)$ and $F \cap \overline{G}_0 = \emptyset$. There is a representation

$$x = x_0 + x_1,\quad x_0 \in E(G_0),\quad x_1 \in E(G).$$

For $\lambda \in W = F^c \cap (\overline{G}_0)^c \subset \rho(x,T) \cap \rho(x_0,T) \cap \rho(x_1,T)$, we have

$$x(\lambda) = x_0(\lambda) + x_1(\lambda).$$

Since $E(G)$ is a μ-space of T, $\{x_1(\lambda) : \lambda \in W\} \subset E(G)$. Let Δ be a Cauchy neighborhood of F such that $\overline{\Delta} \cap \overline{G}_0 = \emptyset$ and $\Gamma = \partial\Delta \subset W$. Then

$$x = \frac{1}{2\pi i}\int_\Gamma x(\lambda)d\lambda = \frac{1}{2\pi i}\int_\Gamma x_0(\lambda)d\lambda$$

$$+ \frac{1}{2\pi i}\int_\Gamma x_1(\lambda)d\lambda = \frac{1}{2\pi i}\int_\Gamma x_1(\lambda)d\lambda \in E(G)$$

and hence (13.10) follows.

(III) => (IV): Let $\lambda \in G$ and choose $\{G_0,G_1\} \in \text{cov}\, \sigma(T)$ with $G_0 \in V_\infty$, $G_1 \in G^K$, $\overline{G}_1 \subset G$ and $\lambda \notin \overline{G}_0$. By 11.2 and (13.10), we have

$$X = E(G_0) + E(G_1) = E(G_0) + \Xi(T,\overline{G}_1) = E(G_0) + E(G).$$

For $X_0 = E(G_0)$, $X_1 = \Xi(T,\bar{G}_1)$ and $Y = E(G)$, 3.2 implies that $\hat{T} = T/E(G)$ is closed and $\sigma(\hat{T}) = \sigma(\tilde{T})$, where $\tilde{T} = [T|E(G_0)]/E(G) \cap E(G_0)$. Since $\lambda \notin \bar{G}_0$, (13.9) implies that $\lambda \notin \sigma[T|E(G_0) \cap E(G)] \cup \sigma[T|E(G_0)]$. Thus $\lambda \in \rho(\hat{T}) = \rho(\tilde{T})$ and since $\lambda \in G$ is arbitrary, $\sigma(\hat{T}) \subset G^c$ is obtained.

(IV) => (I): Put $Y = E(G)$. Then $\text{Int}\,\sigma(T|Y) \subset G$ and the hypothesis implies

$$\sigma(\hat{T}) \cap \text{Int}\,\sigma(T|Y) = \emptyset.$$

Thus Y is analytically invariant under T, by 13.1. □

For $G \in G_\alpha$, (13.10) is no longer independent but a consequence of (13.9).

13.8. THEOREM. Let T have a spectral resolvent E. For every $G \in G_\alpha$, the following assertions are equivalent:

(I) $E(G)$ is analytically invariant under T;

(II) $E(G)$ is a μ-space of T;

(III) for every $H \in G$,

(13.12) $$\sigma[T|E(G) \cap E(H)] \subset \bar{H};$$

(IV) the coinduced $\hat{T} = T/E(G)$ is
(a) closed, if $G \in G^K$;
(b) bounded, if $G \in V_\infty$;
and in both cases, $\sigma(\hat{T}) \subset G^c$.

PROOF. In view of the analogy between this theorem and the foregoing one, the only segment that needs proof is that (13.9) implies (13.10). We confine the proof to $G \in V_\infty$. Let $F \subset G$ be compact and $x \in \Xi(T,F)$ be arbitrary. Choose a relatively compact open H such that $\{G,H\} \in \text{cov}\,\sigma(T)$ and $\bar{H} \cap F = \emptyset$. There is a representation

(13.13) $$x = x_1 + x_2, \quad x_1 \in E(G), \quad x_2 \in E(H).$$

The following inclusions are satisfied by the local spectra of x, x_1, x_2:

$$\sigma(x,T) \subset F, \quad \sigma(x_1,T) \subset F \cup (\bar{G} \cap \bar{H}), \quad \sigma(x_2,T) \subset \bar{G} \cap \bar{H}.$$

Consequently, for $\lambda \in F^c \cap (\bar{H})^c \subset \rho(x,T) \cap \rho(x_1,T) \cap \rho(x_2,T)$, we have

$$x(\lambda) = x_1(\lambda) + x_2(\lambda).$$

Let Δ be a Cauchy neighborhood of $\bar{H}$ such that $\bar{\Delta} \cap F = \emptyset$. Then

$\partial\Delta \subset F^c \cap (\overline{H})^c$ and

(13.14) $$x_2 = \frac{1}{2\pi i}\int_{\partial\Delta} x_2(\lambda)d\lambda = \frac{1}{2\pi i}\int_{\partial\Delta} x(\lambda)d\lambda - \frac{1}{2\pi i}\int_{\partial\Delta} x_1(\lambda)d\lambda = -\frac{1}{2\pi i}\int_{\partial\Delta} x_1(\lambda)d\lambda.$$

By 3.11, for every $\lambda_0 \in \partial\Delta$, there is a neighborhood $V \subset (\overline{H})^c$ of λ_0 and there are functions $f_1 : V \to E(G)$, $f_2 : V \to E(H)$, analytic on V with

$$x_1(\lambda) = f_1(\lambda) + f_2(\lambda) \quad \text{on } V.$$

Then

$$x_1 = (\lambda-T)x_1(\lambda) = (\lambda-T)f_1(\lambda) + (\lambda-T)f_2(\lambda).$$

Since $T|E(H)$ is bounded, the function $g : V \to E(G) \cap E(H)$, defined by

$$g(\lambda) = x_1 - (\lambda-T)f_1(\lambda) = (\lambda-T)f_2(\lambda) = [\lambda-T|E(H)]f_2(\lambda),$$

is analytic on V. In view of (13.12), $V \subset (\overline{H})^c \subset \rho[T|E(G) \cap E(H)]$ and hence the function $h : V \to E(G) \cap E(H)$, defined by

$$h(\lambda) = R[\lambda;T|E(G) \cap E(H)]g(\lambda),$$

is analytic on V. Since T has the SVEP, it follows from

$$(\lambda-T)[f_2(\lambda) - h(\lambda)] = 0$$

that

$$f_2(\lambda) = h(\lambda) \in E(G) \cap E(H) \subset E(G) \quad \text{on } V.$$

Thus $x_1(\lambda_0) \in E(G)$ and since λ_0 is arbitrary on $\partial\Delta$, (13.14) implies that $x_2 \in E(G)$. Then $x \in E(G)$, by (13.13) and hence $\Xi(T,F) \subset E(G)$. □

For any closed $F \subset G$, $X(T,F) \subset E(G)$ has an identical proof.

Summarizing the results of 13.7 and 13.8, we have

13.9. THEOREM. Given T with a spectral resolvent E, the following assertions are equivalent:

(i) E is analytically invariant under T;

(ii) For every $G \in \mathcal{G}$, $E(G)$ is a μ-space of T;

(iii) (a) for every pair $G_1, G_2 \in \mathcal{G}$, $\sigma[T|E(G_1) \cap E(G_2)] \subset \overline{G}_1 \cap \overline{G}_2$;

(b) for every $G \in \mathcal{G}_\beta$ and compact $F \subset G$, $\Xi(T,F) \subset E(G)$;

(iv) For every $G \in G$, $T/E(G)$ is

(1) closed, if $G \notin V_\infty$, (2) bounded, if $G \in V_\infty$,

and in both cases (1), (2), $\sigma[T/E(G)] \subset G^c$.

Next, we shall obtain another characterization of the analytic invariance of $E(\cdot)$, in terms of a "localized" version of the T-absorbent spaces.

13.10. DEFINITION. Given $G \in G$, $Y \in \text{Inv } T$ is said to be T-*absorbent in* G if, for any $\lambda \in G \cap \sigma(T|Y)$ and each $y \in Y$, $(\lambda-T)x = y$ implies that $x \in Y$.

Evidently, if $G \supset \sigma(T|Y)$, then Y is a T-absorbent space.

13.11. THEOREM. Given T with a spectral resolvent E, let $G \in G^K$. The following assertions are equivalent.

(i) $E(G)$ is analytically invariant under T;

(ii) $E(G)$ is T-absorbent in G and $\hat{T} = T/E(G)$ is closed.

PROOF. (i) => (ii): Let $x \in X$ be a solution of equation

$$(\lambda-T)x = y \quad \text{for} \quad \lambda \in G \cap \sigma[T|E(G)] \quad \text{and} \quad y \in E(G).$$

On $X/E(G)$, there corresponds

(13.15) $$(\lambda-\hat{T})\hat{x} = \hat{0} \quad \text{for} \quad \lambda \in G \cap \sigma[T|E(G)].$$

By 13.8, $\hat{T}$ is closed and $\sigma(\hat{T}) \subset G^c$. Then, (13.15) implies that $\hat{x} = \hat{0}$ and hence $x \in E(G)$. Therefore, $E(G)$ is T-absorbent in G.

(ii) => (i): Let $f : \omega_f \to \mathcal{D}_T$ be analytic and satisfy condition

$$(\lambda-T)f(\lambda) \in E(G) \quad \text{on an open } \omega_f.$$

We assume that ω_f is connected. Then $(\lambda-\hat{T})\hat{f}(\lambda) = \hat{0}$ and, since $\hat{T}$ is closed, $E(G) \subset \mathcal{D}_T$. Lemma 4.15, applied to $Y = E(G)$, asserts the existence of an analytic function $h : \omega_h(\subset\omega_f) \to \mathcal{D}_T$ such that $\hat{h}(\lambda) = \hat{f}(\lambda)$ and $(\lambda-T)h(\lambda)$ is analytic on ω_h. If

$$\omega_h - \{G \cap \sigma[T|E(G)]\} = (\omega_h - G) \cup \{\omega_h - \sigma[T|E(G)]\} \neq \emptyset$$

then, at least, one of the sets $\omega_h - G$, $\omega_h - \sigma[T|E(G)]$ is nonvoid. Note that if $\omega_h - G \neq \emptyset$, then $\omega_h - \overline{G} \neq \emptyset$. Let ω denote any of the sets $\omega_h - \overline{G}$, $\omega_h - \sigma[T|E(G)]$, which is assumed to be nonvoid. Then

$$\omega \subset \rho[T|E(G)].$$

The function $g : \omega_h \to \mathcal{D}_T$, defined by $g(\lambda) = (\lambda-T)h(\lambda)$ is analytic on ω_h. It follows from

$$\hat{g}(\lambda) = (\lambda-\hat{T})\hat{h}(\lambda) = (\lambda-\hat{T})\hat{f}(\lambda) = \hat{0}$$

that $g(\lambda) \in E(G)$. For $\lambda \in \omega$, we have

$$(\lambda-T)\{h(\lambda) - R[\lambda;T|E(G)]g(\lambda)\} = 0$$

and hence

$$h(\lambda) = R[\lambda;T|E(G)]g(\lambda) \in E(G) \quad \text{on} \quad \omega.$$

Then $\hat{h}(\lambda) = \hat{f}(\lambda)$ implies that $f(\lambda) \in E(G)$ on ω and $f(\lambda) \in E(G)$ on ω_f, by analytic continuation.

Next, assume that $\omega_h \subset G \cap \sigma[T|E(G)]$. Then, by the hypothesis on E(G),

$$(\lambda-T)h(\lambda) \in E(G)$$

implies that $h(\lambda) \in E(G)$ on ω_h. Thus, $f(\lambda) \in E(G)$ on ω_h and $f(\lambda) \in E(G)$ on ω_f, by analytic continuation. □

If $T \in B(X)$, then in 13.9, condition (iii,b) is redundant and (iv) can be simplified. Furthermore, another characterization of the analytic invariance of E can be expressed in terms of the conjugate T*.

13.12. COROLLARY. Let $T \in B(X)$ have a spectral resolvent E. The following assertions are equivalent:

(I) E is analytically invariant under T;

(II) For every $G \in \mathcal{G}$, E(G) is a μ-space of T;

(III) For every $G \in \mathcal{G}$, E(G) is T-absorbent in G;

(IV) For every pair $G_1, G_2 \in \mathcal{G}$, $\sigma[T|E(G_1) \cap E(G_2)] \subset \bar{G}_1 \cap \bar{G}_2$;

(V) For every $G \in \mathcal{G}$, $\sigma[T/E(G)] \subset G^c$;

(VI) For every $G \in \mathcal{G}$, $\sigma[T^*|E(G)^a] \subset G^c$.

§.14. SPECTRAL CAPACITIES.

Spectral capacities play a role in the theory of decomposable operators, analogous to the role of spectral resolvents in the structure of operators with the SDP. The spectral capacity concept, introduced by Apostol [Ap.1968,c], subsequently evolved to an auxiliary for bounded decomposable operators [Fo.1968].

During our brief thrust into this circle of ideas, we shall both weaken and strengthen the spectral capacity concept for the benefit of gaining a further insight into the spectral theoretic structure of the given operator.

14.1. DEFINITION. Given $n \in \mathbb{N}$, a map $\eta : \mathcal{F} \to \mathcal{S}(X)$ is called a n-*prespectral capacity* if it satisfies the following conditions:

(I) $\{F_k\}_{k=1}^{\infty} \subset \mathcal{F}$ implies $\eta(\bigcap_{k=1}^{\infty} F_k) = \bigcap_{k=1}^{\infty} \eta(F_k)$;

(II) for every $\{G_i\}_{i=0}^{n} \in \text{cov}\, \mathbb{C}$ with $G_0 \in V_\infty$, $\{G_i\}_{i=1}^{n} \subset G^K$,

$$X = \sum_{i=0}^{n} \eta(\overline{G}_i).$$

Note that (II) implies $\eta(\mathbb{C}) = X$.

Given T, η is said to be a n-*prespectral capacity of* T if η satisfies conditions (I), (II) above and

(III) $\eta(F) \in \text{Inv}\, T$, $\sigma[T|\eta(F)] \subset F$ for all $F \in \mathcal{F}$.

The set function η is a n-*spectral capacity of* T if it satisfies conditions (I) - (III) above and

(IV) $\eta(F) \subset \mathcal{D}_T$ if $F \in \mathcal{F}^K$.

If, for every $n \in \mathbb{N}$, η is a n- (pre-) spectral capacity of T, then η is a (pre-) *spectral capacity of* T.

14.2. REMARKS. (a) Condition (I) can be replaced by

(I') for every collection $\mathcal{F}' \subset \mathcal{F}$, $\eta(\bigcap_{F \in \mathcal{F}'} F) = \bigcap_{F \in \mathcal{F}'} \eta(F)$;

(b) if η is a n-prespectral capacity of T then $\eta(\emptyset) \subset Z$, where Z is the set defined by 11.5. Moreover, if η is a n-spectral capacity of T then $\eta(\emptyset) = \{0\}$.

14.3. THEOREM. Given T, the following assertions are equivalent:

(i) T has the n-SDP;

(ii) T has a n-prespectral capacity η such that $\eta(\emptyset)$ is a spectral maximal space of T. Moreover, η is unique.

PROOF. (ii) => (i): Let $F \in F^K$. In view of condition (III), 3.9 gives the direct sum

(14.1) $$\eta(F) = \eta'(F) \oplus W$$

where $\eta'(F) \subset \mathcal{D}_T$, $\sigma[T|\eta'(F)] = \sigma[T|\eta(F)] \subset F$ and $\sigma(T|W) = \emptyset$. Since $\eta(\emptyset) \in SM(T)$, $W \subset \eta(\emptyset)$. For $G \in G$, put

(14.2) $$E(G) = \begin{cases} \eta'(\overline{G}), & \text{if } G \in G^K; \\ \eta(\overline{G}), & \text{if } G \text{ is unbounded.} \end{cases}$$

It follows from 14.1 (I) that, for each $G \in G$, $\eta(\emptyset) \subset \eta(\overline{G})$ and hence (14.2) implies that $\eta(\emptyset) \subset E(G)$, for each unbounded $G \in G$. Thus, by 14.1 (II), for any $\{G_i\}_{i=0}^n \in \text{cov}\ \mathbb{C}$ with $G_0 \in V_\infty$, $\{G_i\}_{i=1}^n \subset G^K$, we have

$$X = \sum_{i=0}^n \eta(\overline{G}_i) = \eta(\overline{G}_0) + \sum_{i=1}^n \eta'(\overline{G}_i) = \sum_{i=0}^n E(G_i).$$

In view of 14.1 (III), T has the n-SDP. Note that E, as defined by (14.2), is a n-spectral resolvent.

To see that η is unique, note that for every $F \in F$, the members of G_F (11.2) are unbounded. Apply 14.2 (a), (14.2) and 11.3 to infer that

$$X(T,F) = \bigcap_{G \in G_F} E(G) = \bigcap_{G \in G_F} \eta(\overline{G}) = \eta(\bigcap_{G \in G_F} \overline{G}) = \eta(F).$$

(i) => (ii): In view of 5.9, we can define

$$\eta(F) = X(T,F), \quad F \in F.$$

By the properties of $X(T,\cdot)$, η is a n-prespectral capacity of T. Since $\eta(\emptyset) = X(T,\emptyset)$ is a spectral maximal space of T, (ii) holds. □

14.4. COROLLARY. If η is a 1-prespectral capacity of T such that $\eta(\emptyset) \in SM(T)$ then, for each $n \in \mathbb{N}$, η is a n-prespectral capacity of T and hence a prespectral capacity of T.

PROOF. Let η have the properties stated by the corollary. T has the 1-SDP by 14.3. Then, for every $n \in \mathbb{N}$, T has the n-SDP by 6.3. Quoting again 14.3, η is a n-prespectral capacity of T and hence a prespectral capacity of T. □

Next, we extend an important characterization of bounded decomposable operators to the unbounded case.

14.5. COROLLARY. T is decomposable iff T has a spectral capacity η. In this case, η is unique.

PROOF. T is decomposable iff it has the SDP and $X(T,\emptyset) = \{0\}$, by 6.4. In this case, T has a unique prespectral capacity, by 14.3. For each $F \in F^K$,

$$\eta(F) = \eta'(F) \oplus W \subset \eta'(F) + \eta(\emptyset) = \eta'(F) + \{0\} = \eta'(F) \subset \mathcal{D}_T$$

and hence η is a spectral capacity of T, by 14.1 (IV).

Conversely, if T has a spectral capacity η, then $\eta(\emptyset) = \{0\}$, by 14.2 (b). Furthermore, 11.3 can be expressed in terms of η instead of E, as follows

$$X(T,F) = \bigcap_{G \in G_F} \eta(\overline{G}).$$

Then, for $F = \emptyset$, we obtain

$$X(T,\emptyset) = \bigcap \{\eta(\overline{G}) : G \in G \cap V_\infty\}$$
$$= \eta(\bigcap \overline{G} : G \in G \cap V_\infty\} = \eta(\emptyset) = \{0\}.$$

Thus, it follows from 6.4 that T is decomposable. By 14.3, η is unique. □

In [E.1975] an extension of the decomposable operator concept to the unbounded case was obtained, by means of a strong version of the spectral capacity. An extra feature of the strong spectral capacity η is that, for any $F \in F$, the linear manifold

(14.3) $$M(F) = \{x \in \eta(C) : C \in F^K, C \subset F\}$$

is dense in $\eta(F)$. In this vein, we have the following

14.6. THEOREM. Let T have a spectral capacity η. Then

(i) $\overline{M(\mathbb{C})} = X$ iff T is densely defined and T* is decomposable;

(ii) for every closed $F \in V_\infty$, $\overline{M(F)} = \eta(F)$ if T is densely defined and T* is decomposable.

PROOF. By 14.5, T is decomposable and $\eta(F) = X(T,F)$, $F \in F$.

(i): In particular, T has the SDP and hence T* has the SDP, by the duality theorem 8.1.

Let $\{G_n\}_{n=0}^\infty \subset G^K$ be such that $G_0 = \emptyset$, $\overline{G}_n \subset G_{n+1}$ $(n \in \mathbb{N})$

and $\bigcup_{n=1}^{\infty} G_n = \mathbb{C}$. Putting $F_n = G_n^c$ $(n \in \mathbb{N})$, we have $\bigcap_{n=1}^{\infty} F_n = \emptyset$. Then $X(T,\overline{G}_n) \subset X(T,G_{n+1})$ and since T is decomposable, $\Xi(T,G_n) = X(T,G_n)$, $n \in Z^+$. Assume that T is densely defined, apply 9.8 and obtain successively:

$$X^*(T^*,\emptyset) = \bigcap_{n=1}^{\infty} X^*(T^*,F_n) = [\bigcup_{n=1}^{\infty} \overline{\Xi(T,G_n)}]^a$$

$$= [\bigcup_{n=1}^{\infty} \overline{X(T,G_n)}]^a = [\bigcup_{n=0}^{\infty} X(T,\overline{G}_n)]^a = [\bigcup_{n=1}^{\infty} \eta(\overline{G}_n)]^a.$$

Property (i) now follows from the sequel of the equivalent statements:

(a) T^* is decomposable; (b) $X^*(T^*,\emptyset) = \{0\}$;

(c) $[\bigcup_{n=1}^{\infty} \eta(\overline{G}_n)]^a = \{0\}$; (d) $\overline{\bigcup_{n=1}^{\infty} \eta(\overline{G}_n)} = X$.

(ii): Assume that T is densely defined and T^* is decomposable. Let $F \in V_\infty$ be closed, put $H_1 = \text{Int } F$ and choose $H_2 \in G^K$ such that $\mathbb{C} = H_1 \cup H_2$. Let $x \in X(T,F)$. T being decomposable,

$$X = X(T,\overline{H}_1) + X(T,\overline{H}_2). \tag{14.4}$$

Since, by (i), $\overline{M(\mathbb{C})} = X$, there is a sequence $\{x_n\} \subset \mathcal{D}_T$ with $\sigma(x_n,T)$ compact for each n, and $x_n \to x$. In view of (14.4), there is a representation

$$x - x_n = x_{n1} + x_{n2}, \quad x_{ni} \in X(T,\overline{H}_i), \quad i=1,2$$

and there is a constant $M > 0$ such that

$$\|x_{n1}\| + \|x_{n2}\| \le M \|x - x_n\|.$$

Then, $\|x - x_n\| \to 0$ implies that $\|x_{ni}\| \to 0$ $(i=1,2)$. For the vector

$$y_n = x - x_{n1} = x_n + x_{n2}, \quad n \in \mathbb{N}$$

we have $\sigma(y_n,T) \subset F \cap [\sigma(x_n,T) \cup \overline{H}_2]$. Since $\sigma(x_n,T) \cup \overline{H}_2$ is compact, $y_n \in M(F)$. Consequently, $\|x - y_n\| = \|x_{n1}\| \to 0$ implies that $\overline{M(F)} = \eta(F)$. □

NOTES AND COMMENTS

The spectral resolvent concept first appeared in [E.1980], under the name of spectral resolution, and subsequently was studied in [E.1979] and [Sh.1979] (the apparent date-sequel conflict of the above mentioned publications is technical). Its extension to unbounded closed operators was first applied to analytically invariant spectral resolvents in [W-E.1984].

Lemma 11.2 (iii) generalizes a property of the spectral resolvent pertinent to a bounded operator [Sh.1979, Proposition 16]. The proof of 11.3 is based on a technique employed in [N.1978]. The prespectral resolvent concept was introduced in [W-E.1984,a].

Section 12 is based on [E-W.1983]. In [E.1981], monotonic properties of a spectral resolvent E resulted in two cases: (a) the superset G_2 was the complement of a closed disk; (b) the subset G_1 was an open disk with the range of E in the collection of the T-absorbent subspaces.

Section 13 contains results published in [W-E.1984]. In an earlier paper [W-E.1983,a], the analytically invariant spectral resolvent has been defined and studied for bounded operators. In such a case, there are some interesting implications between various spectral properties of E and some special types of spectral resolvents, such as monotonic, almost localized, analytically invariant and strong spectral resolvents. To present a schematic diagram of the implications, we direct the reader onward to 16.1 and 16.2, and give names to some spectral properties of E. We say that E has the *spectral invariance property* (SIP) if, for any pair $G_1,G_2 \in G$, the inclusion

$$\sigma[T|E(G_1) \cap E(G_2)] \subset \overline{G}_1 \cap \overline{G}_2$$

holds. We say that E has the *inclusion property* (IP) if, for every $G \in G$, one has

$$\overline{X(T,G)} \subset E(G) \subset X(T,\overline{G}).$$

Given $T \in B(X)$ endowed with a spectral resolvent E, the following *diagram of implications* holds:

E is analytically invariant $\Longleftrightarrow$ E has the SIP $\Longrightarrow$ E has the IP

$\Downarrow$ $\Uparrow$ $\Updownarrow$

E is monotonic $\Longleftarrow$ E is strongly monotonic

$\Uparrow$ $\Updownarrow$

E is a strong spectral resolvent $\Longrightarrow$ E is almost localized

Section 14 is based on [W-E.1984,a]. The n-spectral capacity was introduced in [A-V.1974]. The *strong spectral capacity* concept was used by Albrecht [A.1978,a] in a different context. The map η, defined in 14.1, is a strong spectral capacity, if condition (II) is replaced by

(II') For every $\{G_i\}_{i=0}^{n} \in \text{cov}\,\mathbb{C}$ and any $F \in \mathcal{F}$,

$$\eta(F) = \sum_{i=0}^{n} \eta(F \cap \overline{G}_i).$$

The strong spectral capacity is used to give a characterization to the strongly decomposable operators (the latter will be studied in a different framework, in the next section).

The *support of a spectral capacity* η was introduced in [Ap.1968,c] by

$$\text{Supp}\,\eta = \bigcap \{F \in \mathcal{F} : \eta(F) = X\}.$$

By the covering theorem of Lindelöf, Supp η can be expressed by a countable intersection. For more on the support see [Ev.1981].

In [A-V.1974], the concepts of spectral and strong spectral capacities have been extended to finite systems of commuting bounded operators in the sense of [Tay.1970]. More on this is given in [V.1982, Chapter IV. Sect.1].

CHAPTER V. SPECIAL TOPICS IN SPECTRAL DECOMPOSITION

Some spectral decompositions of a special nature form the subject of this chapter.

§.15. THE STRONG SPECTRAL DECOMPOSITION PROPERTY.

An important question about linear operators concerns their ability to transmit the spectral properties to restrictions and coinduced operators. More to the point, is the SDP of an operator transmissible to its restrictions? The answer is negative even for bounded decomposable operators [A.1978,a]. By strengthening the SDP to induce that property to certain restrictions of the given operator, we may expect richer spectral properties. In the bounded case, the concept of *strongly decomposable operator*, introduced in [Ap.1968,a], fulfills the expectations. In this section, we extend this concept to unbounded operators and subsequently, we shall come back to strongly decomposable operators with some new properties.

15.1. DEFINITION. T is said to have the *strong spectral decomposition property* (SSDP) if, for every spectral maximal space Y of T and any $\{G_i\}_{i=0}^n \in \text{cov}\,\sigma(T)$ with $G_0 \in V_\infty$ and $\{G_i\}_{i=1}^n \subset G^K$, there exists $\{X_i\}_{i=0}^n \subset \text{Inv}\,T$ satisfying the following conditions:

(I) $X_i \subset \mathcal{D}_T$ if $G_i \in G^K$ $(1 \leq i \leq n)$;

(II) $\sigma(T|X_i) \subset G_i$, $0 \leq i \leq n$;

(III) $Y = \sum_{i=0}^{n} Y \cap X_i$.

Clearly, the SSDP implies the SDP. Consequently, if T is endowed with the SSDP then it has property (β) and, in particular, enjoys property (κ). Moreover, the decomposition (III) can be expressed in terms of spectral maximal and T-bounded spectral maximal spaces:

$$Y = Y \cap X(T,\overline{G}_0) + \sum_{i=1}^{n} Y \cap \Xi(T,\overline{G}_i). \tag{15.1}$$

15.2. LEMMA. Let T have property (κ). For every compact F, $T|X(T,F)$ has the SDP iff $T|\Xi(T,F)$ is decomposable.

PROOF. Denote $Y = X(T,F)$, $Z = \Xi(T,F)$ and quote 4.34 to write

$$Y = Z \oplus X(T,\emptyset). \tag{15.2}$$

Let $\{G_0,G_1\} \in \operatorname{cov} \sigma(T|Y) = \operatorname{cov} \sigma(T|Z)$ with $G_0 \in V_\infty$ and $G_1 \in G^K$. If $T|Y$ has the SDP then Y admits a spectral decomposition of the type (5.8):

$$Y = Y(T|Y,\overline{G}_0) + \Upsilon(T|Y,\overline{G}_1)$$

where $Y(T|Y,\overline{G}_0) \in SM(T|Y)$ and $\Upsilon(T|Y,\overline{G}_1)$ is the $T|Y$-bounded spectral maximal space associated to $Y(T|Y,\overline{G}_1)$, in the sense of 4.34. It follows from $\Upsilon(T|Y,\overline{G}_1) \subset \Xi(T,F) = Z$ and from 4.25 that

$$Z = Z \cap Y(T|Y,\overline{G}_0) + \Upsilon(T|Y,\overline{G}_1)$$

and hence $T|Z$ is decomposable.

Conversely, if $T|Z$ is decomposable, then

$$Z = Z(T|Z,\overline{G}_0) + Z(T|Z,\overline{G}_1)$$

and it follows from (15.2) that

$$Y = [Z(T|Z,\overline{G}_0) \oplus X(T,\emptyset)] + Z(T|Z,\overline{G}_1).$$

Thus $T|Y$ has the SDP. □

15.3. LEMMA. Given T with the SSDP, let $Y \in SM(T)$, (or $Y \in SM_b(T)$). If $\hat{Z} \in SM(\hat{T})$, where $\hat{T} = T/Y$, then $Z = \{x \in X : x \in \hat{x},\ \hat{x} \in \hat{Z}\}$ is a spectral maximal space of T.

PROOF. $\hat{T}$ is closed by 5.16 (we may assume that $\sigma(T|Y) \neq \mathbb{C}$). The assertion of the lemma will follow by showing that

$$Z = X[T,\sigma(T|Z)]. \tag{15.3}$$

The inclusion $Z \subset X[T,\sigma(T|Z)]$ is evident (e.g. (4.21)). Let $Y \in SM(T)$. To obtain the opposite inclusion put $W = X[T, \sigma(T|Z)]$, apply 5.16 and write

$$\sigma[(T|W)/Y] = \overline{\sigma(T|W) - \sigma(T|Y)} \subset \overline{\sigma(T|Z) - \sigma(T|Y)}. \tag{15.4}$$

Proposition 3.1 applied to $T|Z$ gives

$$\sigma(T|Z) \subset \sigma(T|Y) \cup \sigma[(T|Z)/Y]$$

and then (15.4) becomes

$$\sigma[(T|W)/Y] \subset \sigma[(T|Z)/Y], \quad \text{i.e.} \quad \sigma(\hat{T}|\hat{W}) \subset \sigma(\hat{T}|\hat{Z}).$$

Since $\hat{Z} \in SM(\hat{T})$, $\hat{W} \subset \hat{Z}$ and hence $W \subset Z$. Thus (15.3) is obtained and hence $Z \in SM(T)$, by 4.32. The case of $Y \in SM_b(T)$ has an identical proof. □

Our trust in the SSDP fulfills the expectations:

15.4. THEOREM. T has the SSDP iff, for every $Y \in SM(T)$, $T|Y$ has the SDP.

PROOF. (If): Given $Y \in SM(T)$, let $\{G_i\}_{i=0}^n \in \text{cov}\,\sigma(T)$ with $G_0 \in V_\infty$ and $\{G_i\}_{i=1}^n \subset G^K$. Put $Y_i = X(T,\overline{G}_i)$, $0 \le i \le n$, use 5.19, 5.18, 4.11 and 4.9 to write

(15.5) $$Y(T|Y,\overline{G}_i) = Y \cap X(T,\overline{G}_i) \in SM(T|Y), \quad 0 \le i \le n.$$

By hypothesis and (15.5), we have

(15.6) $$\sum_{i=0}^n Y \cap Y_i = \sum_{i=0}^n Y \cap X(T,\overline{G}_i) = \sum_{i=0}^n Y(T|Y,\overline{G}_i)$$
$$= Y(T|Y,\overline{G}_0) + \sum_{i=1}^n \Upsilon(T|Y,\overline{G}_i) = Y,$$

where, for $1 \le i \le n$, $\Upsilon(T|Y,\overline{G}_i)$ is the $T|Y$-bounded spectral maximal space associated to $Y(T|Y,\overline{G}_i)$. Furthermore, we have

(15.7) $$\sigma[(T|Y)|Y_i] = \sigma(T|Y_i) = \sigma[T|X(T,\overline{G}_i)] \subset \overline{G}_i, \quad 0 \le i \le n.$$

Hence T has the SSDP, by (15.6) and (15.7).

(Only if): Given $Y \in SM(T)$, let $\{G_i\}_{i=0}^n \in \text{cov}\,\mathbb{C}$ with $G_0 \in V_\infty$ and $\{G_i\}_{i=1}^n \subset G^K$. Then

$$X = X(T,\overline{G}_0) + \sum_{i=1}^n \Xi(T,\overline{G}_i).$$

For $Y \in SM(T)$, put

(15.8) $$Y_0 = Y \cap X(T,\overline{G}_0) = X[T,\sigma(T|Y)] \cap X(T,\overline{G}_0) = X[T,\sigma(T|Y) \cap \overline{G}_0];$$

(15.9) $$Y_i = Y \cap \Xi(T,\overline{G}_i) = X[T,\sigma(T|Y)] \cap \Xi(T,\overline{G}_i) = \Xi[T,\sigma(T|Y) \cap \overline{G}_i],$$

$1 \le i \le n$. By the SSDP,

(15.10) $$Y = \sum_{i=0}^n Y_i = X[T,\sigma(T|Y) \cap \overline{G}_0] + \sum_{i=1}^n \Xi[T,\sigma(T|Y) \cap \overline{G}_i].$$

In view of (15.8), (15.9),

(15.11) $$\sigma(T|Y \cap Y_0) = \sigma\{T|X[T,\sigma(T|Y) \cap \overline{G}_0]\} \subset \overline{G}_0;$$

(15.12) $$\sigma(T|Y \cap Y_i) = \sigma\{T|\Xi[T,\sigma(T|Y) \cap \overline{G}_i]\} \subset \overline{G}_i, \quad 1 \le i \le n.$$

Moreover, for $1 \le i \le n$,

(15.13) $$Y_i = Y \cap \Xi(T,\overline{G}_i) \subset \mathcal{D}_{T|Y}.$$

Thus $T|Y$ has the SDP, by 15.10, 15.11, 15.12 and 15.13. □

15.5. THEOREM. T has the SSDP iff the following two conditions hold:

(i) T has the SVEP and $X(T,\emptyset)$ is closed;

(ii) for every $Y \in SM(T)$, $\hat{T} = T/Y$ is closed and $\hat{T}$ has the SSDP.

PROOF. (Only if): (i) is evident. (ii): Given $\{G_i\}_{i=0}^n \in \operatorname{cov} \sigma(\hat{T})$ with $G_0 \in V_\infty$ and $\{G_i\}_{i=1}^n \subset G^K$, let $\hat{Z} \in SM(\hat{T})$. Choose $G_{n+1} \in G^K$ such that $\{G_i\}_{i=0}^{n+1} \in \operatorname{cov} \sigma(T)$ and $G_{n+1} \cap \sigma(\hat{T}) = \emptyset$. Let $\{X_i\}_{i=0}^{n+1} \subset \operatorname{Inv} T$ satisfy conditions:

$$\{X_i\}_{i=1}^{n+1} \subset \mathcal{D}_T; \quad X = \sum_{i=0}^{n+1} X_i; \quad \sigma(T|X_i) \subset G_i, \quad 0 \le i \le n+1.$$

For $0 \le i \le n+1$, put $\sigma_i = \sigma(T|Y) \cup \sigma(T|X_i)$ and $Y_i = X(T,\sigma_i)$. Without loss of generality, we assume that $\sigma(T|Y) \ne \mathbb{C}$. Then $\hat{T}$ is closed by 5.16 and it follows from 4.25 (iii) that

$$\hat{Y}_i = Y_i/Y \in SM(\hat{T}), \quad 0 \le i \le n+1.$$

For $0 \le i \le n$, apply 5.16 to $T|Y_i$ and write

$$(15.14) \qquad \sigma(\hat{T}|\hat{Y}_i) = \sigma[(T|Y_i)/Y] = \overline{\sigma(T|Y_i) - \sigma(T|Y)} \subset \sigma(T|X_i) \subset G_i.$$

For $i=n+1$, it follows from 3.1 that

$$\sigma(T|X_{n+1}) \subset \sigma(T) \cap G_{n+1} = [\sigma(T|Y) \cup \sigma(\hat{T})] \cap G_{n+1} \subset \sigma(T|Y)$$

and since $Y \in SM(T)$, we have $X_{n+1} \subset Y$. Then

$$Y_{n+1} = X(T,\sigma_{n+1}) = X[T,\sigma(T|Y)] = Y$$

and hence $\hat{Y}_{n+1} = \{\hat{0}\}$. For $0 \le i \le n$,

$$X_i \subset X[T,\sigma(T|X_i)] \subset X(T,\sigma_i) = Y_i.$$

By 15.3, $Z = \{x \in X : x \in \hat{x}, \hat{x} \in \hat{Z}\}$ being a spectral maximal space of T, the SSDP of T produces the decomposition

$$Z = \sum_{i=0}^{n+1} Z \cap X_i = \sum_{i=0}^{n+1} Z \cap Y_i.$$

On the quotient space X/Y there corresponds a decomposition of $\hat{Z}$:

$$(15.15) \qquad \hat{Z} = \sum_{i=0}^{n} \hat{Z} \cap \hat{Y}_i$$

It follows from (15.14) and (15.15) that $\hat{T}$ has the SSDP.

(If): $Y = X(T,\emptyset)$ being closed, by hypothesis, it is a spectral maximal space of T. Also, by hypothesis, $\hat{T} = T/Y$ has the SSDP and hence, for closed F, $\hat{X}(\hat{T},F)$ is closed. For closed F, denote

$$X_F = \{x \in X : x \in \hat{x}, \hat{x} \in \hat{X}(\hat{T},F)\}.$$

For $x \in X(T,F)$, it follows from $\sigma(\hat{x},\hat{T}) \subset \sigma(x,T) \subset F$ that $\hat{x} \in \hat{X}(\hat{T},F)$ and hence $X(T,F) \subset X_F$. Conversely, 3.1 asserts that

$$\sigma(T|X_F) \subset \sigma[\hat{T}|\hat{X}(\hat{T},F)] \cup \sigma(T|Y) = \sigma[\hat{T}|\hat{X}(\hat{T},F)] \subset F$$

and hence $X_F \subset X(T,F)$. Therefore,

$$X(T,F) = \{x \in X : x \in \hat{x}, \hat{x} \in \hat{X}(\hat{T},F)\} \tag{15.16}$$

and hence

$$\hat{X}(\hat{T},F) = X(T,F)/Y. \tag{15.17}$$

In view of (15.16), for closed F, X(T,F) is closed.

Now let $Z \in SM(T)$. $\hat{T}$ being closed by hypothesis, 4.25 (iii) implies that $\hat{Z} = Z/Y$ is a spectral maximal space of $\hat{T}$. Also, by hypothesis, $\hat{T}|\hat{Z}$ has the SDP and hence (15.17) can be applied to the restriction $S = T|Z$. Thus, for closed F, we have

$$\hat{Z}(\hat{S},F) = Z(S,F)/Y. \tag{15.18}$$

To show that T has the SSDP, let $\{G_i\}_{i=0}^n \in \text{cov}\,\sigma(T|Z) = \text{cov}\,\sigma(S)$ with $G_0 \in V_\infty$ and $\{G_i\}_{i=1}^n \subset G^K$. It follows from

$$\sigma(T|Z) = \sigma(S) = \sigma(\hat{S}) \cup \sigma(T|Y) = \sigma(\hat{S}) = \sigma(\hat{T}|\hat{Z}),$$

that $\{G_i\}_{i=0}^n \in \text{cov}\,\sigma(\hat{T}|\hat{Z})$. In view of the SDP of $\hat{T}|\hat{Z} = \hat{S}$, we have

$$\hat{Z} = \sum_{i=0}^{n} \hat{Z}(\hat{S},\overline{G}_i). \tag{15.19}$$

Apply (15.18) to $F = \overline{G}_i$ $(0 \le i \le n)$ and obtain

$$\hat{Z} = \sum_{i=0}^{n} Z(S,\overline{G}_i)/Y = \sum_{i=0}^{n} Z(T|Z,\overline{G}_i)/Y$$

or, equivalently,

$$Z = \sum_{i=0}^{n} Z(T|Z,\overline{G}_i) + Y = Z(T|Z,\overline{G}_0) + \sum_{i=1}^{n} \zeta(T|Z,\overline{G}_i),$$

where $\zeta(T|Z,\overline{G}_i)$ is the $T|Z$- bounded spectral maximal space associated to $Z(T|Z,\overline{G}_i)$, $1 \le i \le n$. Thus $T|Z$ has the SDP and hence T has the SSDP, by 15.4. □

15.6. COROLLARY. T has the SSDP iff, for every T-bounded spectral maximal space Y, $\hat{T} = T/Y$ has the SSDP.

PROOF. (If): Since $\{0\} \in SM_b(T)$, for $Y = \{0\}$, $\hat{T} = T$ and the hypothesis implies that T has the SSDP.

(Only if): is identical to the "only if" part of the proof of the foregoing theorem. □

15.7. THEOREM. Given T, the following assertions are equivalent:

(I) T has the SSDP;

(II) (a) T has property (κ);

(b) for every $Y \in SM_b(T)$, $\hat{T} = T/Y$ is closed and, for every $x \in X$,

$$\sigma(\hat{x},\hat{T}) = \overline{\sigma(x,T) - \sigma(T|Y)}; \tag{15.20}$$

(c) for every $Z \in SM(T)$ and each $G \in \mathcal{G}^K$, $G \cap \sigma(T|Z) \neq \emptyset$ implies $\Xi[T,\overline{G} \cap \sigma(T|Z)] \neq \{0\}$.

PROOF. (I) => (II): (a) is evident.

(b): $\hat{T}$ is closed by 15.5. It follows from 4.35, that

$$\sigma(\hat{x},\hat{T}) \supset \overline{\sigma(x,T) - \sigma(T|Y)}.$$

To obtain the opposite inclusion, for $x \in X$, put $F = \sigma(x,T) \cup \sigma(T|Y)$ and apply 5.16 to $T|X(T,F)$, as follows:

$$\sigma(\hat{x},\hat{T}) \subset \sigma[\hat{T}|X(T,F)/Y] = \sigma\{[T|X(T,F)]/Y\}$$

$$= \overline{\sigma[T|X(T,F)] - \sigma(T|Y)} \subset \overline{F - \sigma(T|Y)} = \overline{\sigma(x,T) - \sigma(T|Y)}.$$

(c): Let $Z \in SM(T)$ and $G \in \mathcal{G}^K$. By hypothesis, $T|Z$ has the SDP. Apply 5.16 to $T|Z$ and obtain

$$\overline{G \cap \sigma(T|Z)} \subset \sigma[(T|Z)|Z(T|Z,\overline{G})] =$$

$$\sigma\{T|X[T,\overline{G} \cap \sigma(T|Z)]\} = \sigma\{T|\Xi[T,\overline{G} \cap \sigma(T|Z)]\}.$$

Thus, if $G \cap \sigma(T|Z) \neq \emptyset$ then $\sigma\{T|\Xi[T,\overline{G} \cap \sigma(T|Z)]\} \neq \emptyset$ and hence

$$\Xi[T,\overline{G} \cap \sigma(T|Z)] \neq \{0\}.$$

(II) => (I): Let F be compact with $\text{Int}\, F = G \neq \emptyset$ and let $Z \in SM(T)$. To prove the inclusion

$$\overline{G \cap \sigma(T|Z)} \subset \sigma[T|\Xi(T,F) \cap Z], \tag{15.21}$$

we may suppose that $G \cap \sigma(T|Z) \neq \emptyset$. Let $\lambda_0 \in G \cap \sigma(T|Z)$ and let $\delta \subset G$ be a neighborhood of λ_0. Then $\Xi[\overline{\delta} \cap \sigma(T|Z)] \neq \{0\}$, by (II,c) and hence

$$\sigma[T|\Xi(\overline{\delta} \cap \sigma(T|Z)] \neq \emptyset.$$

Let $\lambda \in \sigma\{T|\Xi[T,\overline{\delta} \cap \sigma(T|Z)]\} \subset \overline{\delta} \cap \sigma(T|Z)$. Then, it follows from

$$\Xi[T,\overline{\delta} \cap \sigma(T|Z)] \subset \Xi[T,F \cap \sigma(T|Z)]$$

$$= \Xi(T,F) \cap X[T,\sigma(T|Z)] = \Xi(T,F) \cap Z$$

that

$$\sigma\{T|\Xi[T,\overline{\delta} \cap \sigma(T|Z)]\} \subset \sigma[T|\Xi(T,F) \cap Z]$$

and hence $\lambda \in \sigma[T|\Xi(T,F) \cap Z]$. Therefore, $\lambda \in \overline{\delta} \cap \sigma[T|\Xi(T,F) \cap Z]$ and consequently

$$\overline{\delta} \cap \sigma[T|\Xi(T,F) \cap Z] \neq \emptyset.$$

Since δ is an arbitrary neighborhood of λ_0, we have $\lambda_0 \in \sigma[T|\Xi(T,F) \cap Z]$ and, by the choice of λ_0, (15.21) follows.

Finally, with the help of (15.21), we shall show that $T|Z$ has the SDP. Denote $W = \Xi(T,F) \cap Z$, $\tilde{Z} = Z/W$, $\tilde{x} = x + W \in \tilde{Z}$ for $x \in Z$, $\tilde{T} = T/W$, use (15.20) and obtain

$$(15.22) \qquad \sigma(\tilde{x},\tilde{T}) = \overline{\sigma(x,T) - \sigma[T|\Xi(T,F) \cap Z]}$$

$$\subset \overline{\sigma(T|Z) - G \cap \sigma(T|Z)} = \sigma(T|Z) - G \subset G^c.$$

Since $Z \in SM(T)$ and $W \in SM_b(T)$, 4.25 (iv) implies that $\tilde{Z} \in SM_b(\tilde{T})$. In particular, $\tilde{Z}$ is a μ-space of $\tilde{T}$. Then, with the help of 2.7 and (15.22), we obtain

$$(15.23) \qquad \sigma(\tilde{T}|\tilde{Z}) = \cup\{\sigma(\tilde{x},\tilde{T}|\tilde{Z}) : \tilde{x} \in \tilde{Z}\} = \cup\{\sigma(\tilde{x},\tilde{T}) : \tilde{x} \in \tilde{Z}\} \subset G^c = (\text{Int } F)^c.$$

In view of (15.23) and by the evident inclusion

$$\sigma[(T|Z)|\Xi(T,F) \cap Z] = \sigma[T|\Xi(T,F) \cap Z] \subset F,$$

$T|Z$ has the SDP, by 5.17. Therefore, T has the SSDP. □

A slight strengthening of condition (II,b) makes (II,c) redundant in 15.7.

15.8. THEOREM. Given T, the following assertions are equivalent:

(I) T has the SSDP;

(II) (A) T has property (κ);

(B) for every $Y \in SM_b(T)$, T/Y is closed;

(C) for every compact F and each $x \in X$,

$$\sigma(\hat{x},\hat{T}) = \overline{\sigma(x,T) - F}, \tag{15.24}$$

where $\hat{T} = T/\Xi(T,F)$, $\hat{x} = x + \Xi(T,F) \in X/\Xi(T,F)$;

(III) (A) T has property (κ);

(B) for every $Y \in SM_b(T)$, T/Y is closed;

(C) for every closed F_1 and compact F_2,

$$\sigma\{[T/\Xi(T,F_2)]|X(T,F_1 \cup F_2)/\Xi(T,F_2)\} \subset F_1. \tag{15.25}$$

PROOF. (I) => (III): We may assume that $F_1 \neq \mathbb{C}$. $T|X(T,F_1 \cup F_2)$ has the SDP, by 15.4. Let $\{G_1,G_2\} \in \text{cov}\,(F_1 \cup F_2)$ with $G_1 \in V_\infty$, $F_1 \subset G_1$, $G_2 \in G^K$ and $\overline{G}_2 \cap F_1 = \emptyset$. For $x \in X(T,F_1 \cup F_2)$, there is a representation

$$x = x_1 + x_2,$$

$$x_1 \in X(T,F_1 \cup F_2) \cap X(T,\overline{G}_1), \quad x_2 \in X(T,F_1 \cup F_2) \cap \Xi(T,\overline{G}_2).$$

It follows from $\sigma(x_2,T) \subset (F_1 \cup F_2) \cap \overline{G}_2 = F_2 \cap \overline{G}_2 \subset F_2$ and $x_2 \in \Xi(T,\overline{G}_2)$ that $x_2 \in \Xi(T,F_2)$. Let $\lambda_0 \notin \overline{G}_1$. Then $\lambda_0 \in \rho\{T|X[T,(F_1 \cup F_2) \cap \overline{G}_1]\}$ and there is $y \in X[T,(F_1 \cup F_2) \cap \overline{G}_1]$ verifying equation $(\lambda_0-T)y = x_1$. To simplify notations, put $W = \Xi(T,F_2)$. By the natural homomorphism $X \to X/W$, we have

$$(\lambda_0-T/W)\hat{y} = \hat{x}_1 = \hat{x}$$

and hence $\lambda_0 - (T/W)|X(T,F_1 \cup F_2)/W$ is surjective. Since T/W has the SVEP by 4.14, 2.5 implies that $\lambda_0 \in \rho[(T/W)|X(T,F_1 \cup F_2)/W]$. By the choice of λ_0, we have

$$\sigma[(T/W)|X(T,F_1 \cup F_2)/W] \subset \overline{G}_1$$

and since $G_1(\supset F_1)$ is arbitrary, inclusion (15.25) holds.

(III) => (II): Given $x \in X$, let F be compact. Apply (15.25) to $Y = \Xi(T,F)$, $F_1 = \overline{\sigma(x,T) - F}$, $F_2 = F$ and obtain

$$\sigma[(T/Y)|X(T,F_1 \cup F_2)/Y] \subset F_1 = \overline{\sigma(x,T) - F}. \tag{15.26}$$

For $x \in X(T,F_1 \cup F_2)$, $\hat{x} = x + Y \in X(T,F_1 \cup F_2)/Y$ and $\hat{T} = T/Y$, (15.26) implies

$$\sigma(\hat{x},\hat{T}) \subset \sigma[\hat{T}|X(T,F_1 \cup F_2)/Y] \subset \overline{\sigma(x,T) - F}.$$

The opposite inclusion follows from 4.35:

$$\sigma(\hat{x},\hat{T}) \supset \sigma(x,T) - \sigma(T|Y) \supset \overline{\sigma(x,T) - F}$$

and hence (15.24) is obtained.

(II) => (I): In view of 15.7, it suffices to show that, for every $G \in G^K$ and $Z \in SM(T)$,

$$(15.27) \qquad G \cap \sigma(T|Z) \neq \emptyset$$

implies $\Xi[T,\overline{G} \cap \sigma(T|Z)] \neq \{0\}$. Let $G \in G^K$ satisfy (15.27) and suppose to the contrary, that $W = \Xi[T,\overline{G} \cap \sigma(T|Z)] = \{0\}$. Then, for $x \in X$, $\tilde{x} = x + W$ and $\tilde{T} = T/W$, we have

$$(15.28) \qquad \sigma(\tilde{x},\tilde{T}) = \sigma(x,T).$$

In view of (15.28), the hypothesis implies

$$\sigma(x,T) = \sigma(\tilde{x},\tilde{T}) = \overline{\sigma(x,T) - \overline{G} \cap \sigma(T|Z)}$$
$$= \overline{\sigma(x,T) - \overline{G}} \cup \overline{\sigma(x,T) - \sigma(T|Z)}.$$

Let $x \in Z$. Since $\sigma(x,T) \subset \sigma(T|Z)$, we have $\sigma(x,T) = \overline{\sigma(x,T) - \overline{G}}$ and hence $\sigma(x,T) \cap G = \emptyset$. Then, with the help of 2.7, one obtains

$$\sigma(T|Z) \cap G = [\bigcup_{x \in Z} \sigma(x,T|Z)] \cap G$$

$$= [\bigcup_{x \in Z} \sigma(x,T)] \cap G = \bigcup_{x \in Z} [\sigma(x,T) \cap G] = \emptyset.$$

But this contradicts (15.27) and hence

$$\Xi[T,\overline{G} \cap \sigma(T|Z)] \neq \{0\}. \ \square$$

We pursue this study for bounded operators. Since a bounded operator with the SDP is decomposable, $T \in B(X)$ endowed with the SSDP is strongly decomposable in the sense of [Ap.1968,a].

15.9. DEFINITION. $T \in B(X)$ is said to be *strongly decomposable* if, for every spectral maximal space Y of T and any $\{G_i\}_{i=1}^n \in \text{cov}\, \sigma(T)$, there is a system $\{X_i\}_{i=1}^n \subset SM(T)$ such that

$$Y = \sum_{i=1}^n Y \cap X_i; \quad \sigma(T|X_i) \subset G_i, \quad 1 \leq i \leq n.$$

We shall use the symbol "≅" for both the equivalence of two subspaces, under isometric isomorphism, and the unitary equivalence of linear operators. If Y and Z are subspaces of X with $Z \subset Y$, then Y/Z is a subspace of X/Z and we denote the annhilator of Y/Z in $(X/Z)^* \cong Z^a$, by $(Y/Z)^a$.

A variant of 3.5, useful in the forthcoming theory, now follows.

15.10. LEMMA. Let Y and Z be subspaces of X with $Z \subset Y$. Then

$$(Y/Z)^a \cong Y^a. \tag{15.29}$$

Furthermore, if Y and Z are invariant under $T \in B(X)$, then

$$(T/Z)^*|(Y/Z)^a \cong T^*|Y^a. \tag{15.30}$$

PROOF. The map $X/Z \to X/Y$ being a continuous homomorphism with kernel Y/Z, the spaces $(X/Z)/(Y/Z)$ and X/Y are isomorphic. For $x \in X$, denote the corresponding elements in the quotient spaces, as follows: $\hat{x} \in X/Y$, $\tilde{x} \in X/Z$, $\mathring{x} \in (X/Z)/(Y/Z)$. Note that $u \in \hat{x}$ iff $u - x \in Y$ iff $(u-x)^{\sim} \in Y/Z$ iff $\tilde{u} \in \mathring{x}$. Since $\inf\{\|v\| : v \in \tilde{u}\} < \|u\|$, we have

$$\begin{aligned} \|\mathring{x}\| = \inf\{\|\tilde{u}\| : \tilde{u} \in \mathring{x}\} &= \inf\inf\{\|v\| : v \in \tilde{u},\ \tilde{u} \in \mathring{x}\} \\ &\le \inf\{\|u\| : u \in \hat{x}\} = \|\hat{x}\|. \end{aligned} \tag{15.31}$$

On the other hand, for every $u \in \hat{x}$, we have $\tilde{u} = u + Z \subset u + Y = \hat{x}$ and hence $\tilde{u} \subset \hat{x}$. Therefore,

$$\inf\{\|v\| : v \in \tilde{u}\} \ge \|\hat{x}\|$$

and this implies

$$\|\mathring{x}\| = \inf\inf\{\|v\| : v \in \tilde{u},\ \tilde{u} \in \mathring{x}\} \ge \|\hat{x}\|. \tag{15.32}$$

Thus, by (15.31) and (15.32), we have $\|\mathring{x}\| = \|\hat{x}\|$. It follows from the equivalences

$$(X/Y)^* \cong Y^a, \quad [(X/Z)/(Y/Z)]^* \cong (Y/Z)^a$$

that (15.29) and (15.30) hold. □

15.11. LEMMA. If $T \in B(X)$ is decomposable then, for every $G \in \mathcal{G}$,

$$X(T,G^c)^a = \overline{X^*(T^*,G)}^w, \tag{15.33}$$

where $\overline{}^w$ is the weak*-closure in X^*.

PROOF. Let $T \in B(X)$ be decomposable. By 10.9, for each closed F,

$$JX(T,F) = X^{**}(T^{**},F) \cap JX, \tag{15.34}$$

where J is the canonical embedding of X into X^{**}. Since T is decomposable, so is T^*. Then, 9.8 applied to T^*, gives

(15.35) $$X^{**}(T^{**},F) = X^*(T^*,F^c)^a.$$

The combination of (15.34) and (15.35) implies

$$X(T,F) = {}^aX^*(T^*,F^c)$$

and hence, for $F = G^c$, (15.33) follows. □

15.12. LEMMA. Given $T \in B(X)$, if T^* is decomposable then, for each $G \in \mathcal{G}$, $\overline{X^*(T^*,G)}^W$ is analytically invariant under T^*.

PROOF. Let $f^* : \omega \to X^*$ be analytic and satisfy condition

$$(\lambda-T^*)f^*(\lambda) \in \overline{X^*(T^*,G)}^W \text{ on an open } \omega.$$

We assume that ω is connected. Put $F = G^c$, $Y = X(T,F)$, use 15.11, 5.17 and obtain

$$\sigma[T^*|\overline{X^*(T^*,G)}^W] = \sigma(T^*|Y^a) = \sigma[(T/Y)^*] = \sigma(T/Y) \subset (\operatorname{Int} F)^c = \overline{G}.$$

First, assume that $\omega \subset \overline{G}$. Then $\omega \subset G \subset \rho(T|Y)$ and, for every $x \in Y$ and $\lambda \in \omega$, we have

$$\langle x,f^*(\lambda)\rangle = \langle(\lambda-T)R(\lambda;T|Y)x,f^*(\lambda)\rangle = \langle R(\lambda;T|Y)x,(\lambda-T^*)f^*(\lambda)\rangle = 0.$$

Since $x \in Y$ is arbitrary, it follows that $f^*(\lambda) \in Y^a = \overline{X^*(T^*,G)}^W$ on ω.

Next, assume that $\omega \not\subset \overline{G}$. Then, the function $h^* : (\omega - \overline{G}) \to X^*$, defined by $h^*(\lambda) = (\lambda-T^*)f^*(\lambda)$, is analytic and gives

(15.36) $$(\lambda-T^*)\{f^*(\lambda) - R[\lambda;T^*|\overline{X^*(T^*,G)}^W]h^*(\lambda)\} = 0.$$

Since T^* has the SVEP,

$$f^*(\lambda) = R[\lambda;T^*|\overline{X^*(T^*,G)}^W]h^*(\lambda) \in \overline{X^*(T^*,G)}^W$$

on $\omega - \overline{G}$ and hence $f^*(\lambda) \in \overline{X^*(T^*,G)}^W$ on ω, by analytic continuation. □

15.13. THEOREM. $T \in B(X)$, (resp. T^*) is strongly decomposable iff

(i) T^* (resp. T) has the SVEP and, for $G \in \mathcal{G}$, $T^*|\overline{X^*(T^*,G)}^W$, (resp. $T|\overline{X(T,G)}$) is decomposable;

(ii) for every pair $G,H \in \mathcal{G}$,

(15.37) $$\overline{X^*(T^*,G\cap H)}^W = \overline{Y^*(T^*|Y^*,H)}^W, \text{ (resp. } \overline{X(T,G\cap H)} = \overline{Y(T|Y,H)}),$$

where $Y^* = \overline{X^*(T^*,G)}^W$, (resp. $Y = \overline{X(T,G)}$).

PROOF. We confine the proof to T, the proof for T* is similar.

(Only if): Let $G \in \mathcal{G}$, $F = G^c$ and $Z = X(T,F)$. The coinduced T/Z is decomposable, by 15.5. In view of 15.11, $X(T,F)^a = \overline{X^*(T^*,G)}^w$ and hence we have

(15.38) $$(X/Z)^* \cong \overline{X^*(T^*,G)}^w.$$

By the duality theorem 8.1, $T^*|\overline{X^*(T^*,G)}^w$ is decomposable. Let $F_1 \supset F$ be closed. With the help of 15.10, the following equivalence is available

(15.39) $$[X(T,F_1)/Z]^a \cong X(T,F_1)^a.$$

Denote $\tilde{T} = T/Z$, $\tilde{X} = X/Z$. Before embarking on the proof of (ii), we shall show that

(15.40) $$\tilde{X}(\tilde{T},\overline{F_1 - F}) = X(T,F_1)/Z.$$

In fact, if $\tilde{x} \in \tilde{X}(\tilde{T},\overline{F_1 - F})$, then $\sigma(\tilde{x},\tilde{T}) \subset \overline{F_1 - F}$ and hence, for each $x \in \tilde{x}$, we have

$$\sigma(x,T) \subset \overline{F_1 - F} \cup F = F_1.$$

Thus $\tilde{x} \in \tilde{X}(\tilde{T},\overline{F_1 - F})$ implies that $x \in X(T,F_1)$ and hence $\tilde{x} \in X(T,F_1)/Z$. Conversely, if $\tilde{x} \in X(T,F_1)/Z = X(T,\overline{F_1 - F} \cup F)/Z$, then 15.8 (III,C) implies

$$\sigma(\tilde{x},\tilde{T}) \subset \sigma[\tilde{T}|X(T,\overline{F_1 - F} \cup F)/Z] \subset \overline{F_1 - F}$$

and hence $\tilde{x} \in \tilde{X}(\tilde{T},\overline{F_1 - F})$. Thus (15.40) is obtained.

Now we are in a position to prove (ii). To simplify notations, put $X^{\bullet} = (\tilde{X})^*$ and $T^{\bullet} = (\tilde{T})^*$. Let $H \in \mathcal{G}$ and put $F_1 = G^c \cup H^c$. Then $F_1 \supset F$ and $\overline{F_1 - F} \subset H^c$. Use 15.11, 15.10, (15.39), (15.38), (15.40) and obtain successively

$$\overline{X^*(T^*,G \cap H)}^w = X(T,F_1)^a \cong [X(T,F_1)/Z]^a$$
$$= \tilde{X}(\tilde{T},\overline{F_1 - F})^a \supset \tilde{X}(\tilde{T},H^c)^a = \overline{X^{\bullet}(T^{\bullet},H)}^w = \overline{Y^*(T^*|Y^*,H)}^w.$$

For the last equality, we used the equivalence

$$T^{\bullet} = [T/X(T,F)]^* \cong T^*|\overline{X^*(T^*,G)}^w = T^*|Y^*.$$

To obtain the opposite inclusion, let $x^* \in X^*(T^*,G \cap H)$. Then

$$\sigma(x^*,T^*) \subset G \cap H \subset G$$

and hence $x^* \in X^*(T^*,G) \subset Y^*$. Since, by 15.12, $Y^* \in AI(T^*)$, we have

$$\sigma(x^*,T^*|Y^*) = \sigma(x^*,T^*) \subset H, \quad \text{i.e.} \quad x^* \in Y^*(T^*|Y^*,H).$$

Thus, it follows that $\overline{X^*(T^*,G \cap H)}^w \subset \overline{Y^*(T^*|Y^*,H)}^w$.

(If): With the notations as in the "only if" part, let F and F_1 be closed. Since $X^*(T^*,\mathbb{C}) = X^*$, (i) implies that T^* is decomposable and hence T is decomposable, by the predual theorem. Therefore $Z = X(T,F)$ is closed. Also $T^*|\overline{X^*(T^*,F^c)}^w$ is decomposable and, by 15.10, we have

$$T^*|\overline{X^*(T^*,F^c)}^w = T^*|X(T,F)^a \cong T^\bullet.$$

Thus $T^\bullet$ is decomposable and hence $\tilde{T}$ is decomposable, by the predual theorem. Consequently, $\tilde{X}(\tilde{T},F_1)$ is closed and

$$\sigma[\tilde{T}|\tilde{X}(\tilde{T},F_1)] \subset F_1. \tag{15.41}$$

Put $G = F^c$, $H = F_1^c$ and $Y^* = \overline{X^*(T^*,G)}^w$. By 15.11, we have

$$T^*|X(T,F \cup F_1)^a = T^*|\overline{X^*(T^*,G \cap H)}^w, \tag{15.42}$$

$$T^\bullet|\tilde{X}(\tilde{T},F_1)^a = T^*|\overline{Y^*(T^*|Y^*,H)}^w. \tag{15.43}$$

It follows from (15.43), (15.37) and (15.42) that

$$T^\bullet|\tilde{X}(\tilde{T},F_1)^a \cong T^*|X(T,F \cup F_1)^a. \tag{15.44}$$

Citing 15.10, we write

$$T^\bullet|[X(T,F \cup F_1)/Z]^a \cong T^*|X(T,F \cup F_1)^a. \tag{15.45}$$

Now, with the help of (15.45), (15.44) and (15.41), we obtain successively

$$\sigma[\tilde{T}|X(T,F \cup F_1)/Z] = \sigma\{T^\bullet|[X(T,F \cup F_1)/Z]^a\}$$
$$= \sigma[T^*|X(T,F \cup F_1)^a] = \sigma[T^\bullet|\tilde{X}(\tilde{T},F_1)^a] = \sigma[\tilde{T}|\tilde{X}(\tilde{T},F_1)] \subset F_1.$$

Thus, conditions (III) of 15.8 are satisfied and hence T is strongly decomposable. □

15.14. THEOREM. Given $T \in B(X)$,

(A) T^* is strongly decomposable iff

(i) T is decomposable;

(ii) for every pair $G,G_1 \in \mathcal{G}$ and any $\varepsilon > 0$,

$$\overline{X(T,G \cup G_1)} \subset \overline{X(T,G_\varepsilon)} + \overline{X(T,G_1)}; \tag{15.46}$$

(B) T is strongly decomposable iff

(i) T^* is decomposable;

(ii) for every pair $G,G_1 \in \mathcal{G}$ and any $\varepsilon > 0$,

$$\overline{X^*(T^*,G \cup G_1)}^w \subset \overline{X^*(T^*,G_\varepsilon)}^w + \overline{X^*(T^*,G_1)}^w,$$

where $G_\varepsilon = \{\lambda : d(\lambda,\overline{G}) < \varepsilon\}$.

PROOF. We confine the proof to (A), that of (B) is similar.

(Only if): In particular, T* is decomposable and so is T, by the predual theorem. Let $G, G_1 \in \mathcal{G}$ be arbitrary and $\varepsilon > 0$. Put $Y = \overline{X(T, G \cup G_1)}$ and let

$$H = \{\lambda : d(\lambda,\overline{G}_1) < \tfrac{\varepsilon}{2}\} \cap \{\lambda : d(\lambda,\overline{G}) > \tfrac{\varepsilon}{2}\}.$$

We have $\overline{G} \cap H = \emptyset$, $\{G_\varepsilon, H\} \in \text{cov}\,(\overline{G} \cup \overline{G}_1)$ and hence $\{G_\varepsilon, H\} \in \text{cov}\,\sigma(T|Y)$. $T|Y$ is decomposable, by 15.13 (i), and hence the following decomposition holds

$$Y = \overline{Y(T|Y, G_\varepsilon)} + \overline{Y(T|Y, H)}\,. \tag{15.47}$$

In view of 15.13 (ii), we have

$$\overline{Y(T|Y, G_\varepsilon)} = \overline{X[T,(G \cup G_1) \cap G_\varepsilon]} \subset \overline{X(T, G_\varepsilon)}, \tag{15.48}$$

$$\overline{Y(T|Y, H)} = \overline{X[T,(G \cup G_1) \cap H]} \subset \overline{X(T, G_1 \cap H)} \subset \overline{X(T, G_1)}. \tag{15.49}$$

Thus, (15.46) follows from (15.47), (15.48) and (15.49).

(If): Since T is decomposable, so is T* and hence $X^*(T^*,F)$ is closed for closed F. In view of 15.4, we only have to show that, for an arbitrary closed F, $T^*|X^*(T^*,F)$ is decomposable.

Let $G_1 = F^c$ and let $G \in \mathcal{G}$ be arbitrary. Put $\hat{X} = X/\overline{X(T,G_1)}$ and $\hat{T} = T/\overline{X(T,G_1)}$. It follows from (15.46) that every $x \in \overline{X(T, G \cup G_1)}$ has a representation

$$x = x_1 + x_\varepsilon, \quad x_1 \in \overline{X(T,G_1)}, \quad x_\varepsilon \in \overline{X(T,G_\varepsilon)}.$$

Consequently, $\hat{x} = \hat{x}_\varepsilon$ and hence

$$\sigma(\hat{x},\hat{T}) = \sigma(\hat{x}_\varepsilon,\hat{T}) \subset \sigma(x_\varepsilon, T) \subset \overline{G}_\varepsilon.$$

Since

$$\hat{Z} = \overline{X(T, G \cup G_1)}/\overline{X(T,G_1)} \tag{15.50}$$

is analytically invariant under $\hat{T}$ (Proposition 4.16 (ii)) and hence a μ-space of $\hat{T}$, we have

$$\sigma(\hat{x},\hat{T}|\hat{Z}) = \sigma(\hat{x},\hat{T}) \subset \overline{G}_\varepsilon. \tag{15.51}$$

With the help of 2.7, it follows from (15.51) that

$$\sigma(\hat{T}|\hat{Z}) = \cup\{\sigma(\hat{x},\hat{T}|\hat{Z}) : \hat{x} \in \hat{Z}\} = \cup\{\sigma(\hat{x},\hat{T}) : \hat{x} \in \hat{Z}\} \subset \overline{G}_\varepsilon.$$

Since $\varepsilon > 0$ is arbitrary, we have

$$\sigma(\hat{T}|\hat{Z}) \subset \overline{G}. \tag{15.52}$$

Now, let $\{H_0,H_1\} \in \text{cov}\,\mathbb{C}$. T being decomposable, we have

$$X = \overline{X(T,H_0)} + \overline{X(T,H_1)} = \overline{X(T,H_0 \cup G_1)} + \overline{X(T,H_1 \cup G_1)}$$

and there corresponds the following decomposition of the quotient space

$$X/\overline{X(T,G_1)} = \overline{X(T,H_0 \cup G_1)}/\overline{X(T,G_1)} + \overline{X(T,H_1 \cup G_1)}/\overline{X(T,G_1)}. \tag{15.53}$$

It follows from (15.52), applied to $G = H_i$, $(i=0,1)$ that

$$\sigma[\hat{T}|\overline{X(T,H_i \cup G_1)}/\overline{X(T,G_1)}] \subset \overline{H}_i, \quad i=0,1. \tag{15.54}$$

$\hat{T}$ is decomposable, by (15.53) and (15.54). It follows from

$$(\hat{T})^* = [T/\overline{X(T,G)}]^* = T^*|X^*(T^*,F),$$

that $T^*|X^*(T^*,F)$ is decomposable, by the duality theorem. □

15.15. PROPOSITION. Given $T \in B(X)$, if T^{***} is strongly decomposable then T^* is strongly decomposable.

PROOF. Assume that T^{***} is strongly decomposable. It follows from 6.5 and from the spectral duality theorems that both T^* and T^{*****} are decomposable. For $Y^* \in SM(T^*)$, use the representation $Y^* = X^*(T^*,F)$ with $F = \sigma(T^*|Y^*)$. Let $\{G_i\}_{i=1}^n \in \text{cov}\,[\sigma(T) = \sigma(T^*) = \sigma(T^{***})]$. Since T^{*****} has the SVEP, relation (10.10) holds and it will be used in the following equalities:

$$\begin{aligned} \sum_{i=1}^n [KY^* \cap KX^*(T^*,\overline{G}_i)] &= \sum_{i=1}^n [KX^*(T^*,F) \cap KX^*(T^*,\overline{G}_i)] \\ &= \sum_{i=1}^n [PX^{***}(T^{***},F) \cap PX^{***}(T^{***},\overline{G}_i)] \\ &= P\sum_{i=1}^n [X^{***}(T^{***},F) \cap X^{***}(T^{***},\overline{G}_i)]. \end{aligned} \tag{15.55}$$

T^{***} being strongly decomposable, the following decomposition holds

$$X^{***}(T^{***},F) = \sum_{i=1}^n [X^{***}(T^{***},F) \cap X^{***}(T^{***},\overline{G}_i)]. \tag{15.56}$$

It follows from (15.55), (15.56) and (10.10) that

$$\sum_{k=1}^n [KY^* \cap KX^*(T^*,\overline{G}_i)] = PX^{***}(T^{***},F) = KX^*(T^*,F) = KY^*$$

which means that $T^{***}|KX^*$ is strongly decomposable and hence so is T^*. □

§.16. STRONG SPECTRAL RESOLVENTS.

A useful property of some operators is that they have decomposable restrictions. A strongly decomposable operator has this property. Is there another way to obtain decomposable restrictions? An affirmative answer arises from a strong version of the spectral resolvent concept.

16.1. DEFINITION. A spectral resolvent E of a closed operator T is said to be a *strong spectral resolvent* of T if, for any $G \in \mathcal{G}$, there is $\{G_i\}_{i=0}^n \in \operatorname{cov} \overline{G}$ with $G_0 \in V_\infty$ and $\{G_i\}_{i=1}^n \subset \mathcal{G}^K$ such that

$$E(G) = \sum_{i=0}^{n} E(G) \cap E(G_i). \tag{16.1}$$

Since (16.1) implies $E(G) \subset \sum_{i=0}^{n} E(G_i)$, every strong spectral resolvent is almost localized.

16.2. THEOREM. Every strong spectral resolvent E of $T \in B(X)$ is analytically invariant under T.

PROOF. In view of 13.9, it suffices to show that, for any $G \in \mathcal{G}$, $E(G)$ is a μ-space of T. Given $G \in \mathcal{G}$, let $x \in E(G)$. For $\lambda_0 \in \rho(x,T)$, there is $r > 0$ such that

$$S(\lambda_0,r) = \{\lambda : |\lambda - \lambda_0| < r\} \subset \rho(x,T).$$

If $\lambda_0 \in (\overline{G})^c$ then $\lambda_0 \in \rho[T|E(G)]$ and $S(\lambda_0,r) \subset \rho[T|E(G)]$, for $r > 0$ sufficiently small. Then

$$x(\lambda) = R[\lambda;T|E(G)]x \in E(G) \quad \text{on} \quad S(\lambda_0,r)$$

and hence $x(\lambda_0) \in E(G)$. If $\lambda_0 \in \partial G$ then, for every $r > 0$, $x(\lambda) \in E(G)$ on $S(\lambda_0,r) \cap (\overline{G})^c$ and $x(\lambda) \in E(G)$ on $S(\lambda_0,r)$, by analytic continuation. Hence $x(\lambda_0) \in E(G)$.

Next, assume that $\lambda_0 \in G$. Then, for $r > 0$ sufficiently small, $S(\lambda_0,r) \subset G$. Put

$$G_1 = \{\lambda : \operatorname{Im} \lambda > \operatorname{Im} \lambda_0 - \tfrac{r}{6}\} - \overline{S(\lambda_0, \tfrac{r}{2})};$$

$$G_2 = \{\lambda : \operatorname{Im} \lambda < \operatorname{Im} \lambda_0 + \tfrac{r}{6}\} - \overline{S(\lambda_0, \tfrac{r}{2})};$$

$$G_3 = S(\lambda_0, \tfrac{2}{3} r).$$

Then $\{G_1,G_2,G_3\} \in \operatorname{cov} \overline{G}$ and, by the hypothesis on E,

$$E(G) = \sum_{i=1}^{3} E(G) \cap E(G_i).$$

There is a representation

$$x = x_1 + x_2 + x_3 \quad \text{with} \quad x_i \in E(G) \cap E(G_i),\ i=1,2,3.$$

By construction, $S(\lambda_0, \frac{r}{3}) \subset \rho[T|E(G) \cap E(G_i)]$, i=1,2 and hence, for $\lambda \in S(\lambda_0, \frac{r}{3})$, we have

$$(\lambda-T)\{x(\lambda) - R[\lambda;T|E(G) \cap E(G_1)]x_1 - R[\lambda;T|E(G) \cap E(G_2)]x_2\}$$
$$= x - x_1 - x_2 = x_3.$$

Consequently,

$$x_3(\lambda) = x(\lambda) - R[\lambda;T|E(G) \cap E(G_1)]x_1 - R[\lambda;T|E(G) \cap E(G_2)]x_2$$

is the local resolvent of x_3. Since E is almost localized, 12.6 and 12.3 imply that $\overline{X(T,G)} \subset E(G)$. Furthermore,

$$\overline{G}_3 = S(\lambda_0, \frac{2r}{3}) \subset S(\lambda_0, r) \subset G$$

and $x_3 \in E(G) \cap E(G_3) \subset E(G_3)$. Then, by 12.4,

$$\{x_3(\lambda) : \lambda \in S(\lambda_0, \frac{r}{3})\} \subset E(G).$$

Thus, for $\lambda \in S(\lambda_0, \frac{r}{3})$,

$$x(\lambda) = R[\lambda;T|E(G) \cap E(G_1)]x_1 + R[\lambda;T|E(G) \cap E(G_2)]x_2 + x_3(\lambda)$$

and hence

$$x(\lambda) = E(G) \cap E(G_1) + E(G) \cap E(G_2) + E(G) = E(G).$$

In particular, $x(\lambda_0) \in E(G)$. Summing up, for any relative location of λ_0 with respect to G, we have

$$\{x(\lambda_0) : \lambda_0 \in \rho(x,T)\} \subset E(G)$$

and hence $E(G)$ is a μ-space of T, by 4.8. □

Our interest in the strong spectral resolvent is motivated by the following

16.3. THEOREM. Let $T \in B(X)$ have a strong spectral resolvent E. Then, for every $G \in \mathcal{G}$, $T|E(G)$ is decomposable.

PROOF. Let $G \in \mathcal{G}$ be given and let $\{G_1, G_2\} \in \text{cov}\, \sigma[T|E(G)]$. If $\overline{G} \not\subset G_1 \cup G_2$, then choose $G_0 \in \mathcal{G}$ such that

$$\overline{G} \subset G_0 \cup G_1 \cup G_2 \quad \text{and} \quad \sigma[T|E(G)] \cap \overline{G}_0 = \emptyset.$$

Then

$$E(G) = \sum_{i=0}^{2} E(G) \cap E(G_i).$$

It follows from 16.2 and 13.12 that

(16.2) $$\sigma[T|E(G) \cap E(G_i)] \subset \overline{G} \cap \overline{G}_i \subset \overline{G}_i, \quad i=1,2$$

and

$$\sigma[T|E(G) \cap E(G_0)] \subset \sigma[T|E(G)] \cap \sigma[T|E(G_0)]$$
$$\subset \sigma[T|E(G)] \cap \overline{G}_0 = \emptyset.$$

Thus $E(G) \cap E(G_0) = \{0\}$ and hence

(16.3) $$E(G) = \sum_{i=1}^{2} E(G) \cap E(G_i).$$

Consequently, $T|E(G)$ is decomposable, by (16.2) and (16.3). □

As an application, we consider a closed operator T whose spectrum lies on $\mathbb{R}$. We say that a set $G \subset \mathbb{R}$, open in the topology of $\mathbb{R}$, is a neighborhood of ∞ in $\mathbb{R}$, in symbols $G \in V_\infty^{\mathbb{R}}$, if for some $r > 0$ sufficiently large, $(-\infty,r) \cup (r,\infty) \subset G$. $\{G_i\}_{i=0}^{n} \in \text{cov}\,\sigma(T) \subset \mathbb{R}$ consists of subsets of $\mathbb{R}$, open in the topology of $\mathbb{R}$ with $G_0 \in V_\infty^{\mathbb{R}}$. Likewise, the domain of a spectral resolvent of T with $\sigma(T) \subset \mathbb{R}$ is the collection $G^{\mathbb{R}}$ of all subsets of $\mathbb{R}$, which are open in the topology of $\mathbb{R}$. Subsets of $\mathbb{C}$ are referred to the topology of $\mathbb{C}$.

16.4. LEMMA. Let T with $\sigma(T) \subset \mathbb{R}$ have a spectral resolvent E. Then, for each $G \in G^{\mathbb{R}}$, $E(G)$ is analytically invariant under T.

PROOF. Given $G \in G^{\mathbb{R}}$, let $f : \omega_f \to \mathcal{D}_T$ be analytic and satisfy condition

$$(\lambda-T)f(\lambda) \in E(G) \quad \text{on an open} \quad \omega_f \subset \mathbb{C}.$$

Since $\sigma(T) \cup \sigma[T|E(G)] \subset \mathbb{R}$, we have

$$\omega_f - \{\sigma(T) \cup \sigma[T|E(G)]\} \neq \emptyset.$$

Put $h(\lambda) = (\lambda-T)f(\lambda)$ and, for $\lambda \in \omega_f - \{\sigma(T) \cup \sigma[T|E(G)]\}$, obtain

$$f(\lambda) = R(\lambda;T)h(\lambda) = R[\lambda;T|E(G)]h(\lambda) \in E(G).$$

Therefore, $f(\lambda) \in E(G)$ on ω_f, by analytic continuation. □

With the method used in the foregoing proof, one can show that, for closed $F \subset \mathbb{R}$ and $G \in G^{\mathbb{R}}$, the subspaces $X(T,F)$ and $\overline{X(T,G)}$ are analytically invariant under T.

16.5. THEOREM. Given T with $\sigma(T) \subset \mathbb{R}$, every spectral resolvent E of T is a strong spectral resolvent.

PROOF. Given $G \in G^{\mathbb{R}}$, let $\{G_i\}_{i=0}^n \in \operatorname{cov} \overline{G}$ with $G_0 \in V_\infty^{\mathbb{R}}$ and $\{G_i\}_{i=1}^n \subset G^{\mathbb{R}}$ relatively compact. The proof consists of: Part A, which builds up a convenient open cover for $\sigma(T)$; Part B, which produces an intermediate spectral decomposition of $E(G)$; and Part C, which establishes the prototype decomposition (16.1) for $E(G)$.

Part A. Define $\{H_j\}_{j=0}^m \in \operatorname{cov} \overline{G}$, as follows:

(I) $H_0 = (-\infty, a_0) \cup (b_0, \infty)$; $H_j = (a_j, b_j)$, $1 \leq j \leq m$ with $a_1 < a_2 < \ldots < a_m$, $a_0 < a_2$, $b_{m-1} < b_0$, $H_j \not\subset H_k$ for $j \neq k$;

(II) $\overline{H}_0 \cup \overline{H}_1 \cup \overline{H}_m \subset G_0$; for each $j = 2,3,\ldots,m-1$ there is an index i $(0 \leq i \leq n)$ such that $\overline{H}_j \subset G_i$.

Since ∂G is zero-dimensional, there are open sets $\{\delta_j\}_{j=0}^m$ with pairwise disjoint closures such that $\{\delta_j\}_{j=0}^m \in \operatorname{cov} \partial G$, $\overline{\delta}_j \subset H_j$ $(0 \leq j \leq m)$. Let $\{\gamma_j\}_{j=0}^m$ be another cover of ∂G with $\overline{\gamma}_j \subset \delta_j$, $0 \leq j \leq m$. Define

$$\Delta_0 = H_0 \cup \Big(\bigcup_{j=0}^m \delta_j\Big) \cup (\overline{G})^c; \quad \Delta_j = \Big(G - \bigcup_{j=0}^m \overline{\gamma}_j\Big) \cap H_j, \quad 1 \leq j \leq m.$$

The collection $\{\Delta_j\}$ has the following properties:

(III) $\Delta_0 \in V_\infty$; $\overline{\Delta}_j \subset G$, $1 \leq j \leq m$;

(IV) $\sigma(T) \subset \bigcup_{j=0}^m \Delta_j$.

To see that (III) holds, for $1 \leq j \leq m$, obtain

$$\overline{\Delta}_j \subset \overline{G - \bigcup_{k=0}^m \overline{\gamma}_k} \subset \overline{G - \bigcup_{k=0}^m \gamma_k} \subset \overline{G} - \bigcup_{k=0}^m \gamma_k \subset \overline{G} - \partial G = G.$$

The proof of (IV) consists in showing that, for every $\lambda_0 \in \sigma(T) - \Delta_0$, there is k $(1 \leq k \leq m)$ such that $\lambda_0 \in \Delta_k$. Indeed, since $\lambda_0 \notin \Delta_0$, we have

$$\text{(16.4)} \qquad \text{(a) } \lambda_0 \notin H_0; \quad \text{(b) } \lambda_0 \notin \bigcup_{j=0}^m \delta_j; \quad \text{(c) } \lambda_0 \in \overline{G}.$$

$\overline{G}$ being covered by $\{H_j\}$, it follows from (16.4,a) that $\lambda_0 \in H_k$, for some $k \neq 0$. Furthermore, (16.4, b and c) imply

$$\lambda_0 \in \overline{G} - \bigcup_{j=0}^m \delta_j \subset G = \bigcup_{j=0}^m \overline{\gamma}_j$$

and hence

$$\lambda_0 \in (G - \bigcup_{j=0}^{m} \overline{\gamma}_j) \cap H_k = \Delta_k, \quad k \neq 0.$$

Part B. In this part, we shall obtain the decomposition

(16.5) $$E(G) = E(G) \cap X(T,\overline{\Delta}_0) + \sum_{j=1}^{m} \Xi(T,\overline{\Delta}_j).$$

The cover $\{\Delta_j\}_{j=0}^{m}$ of $\sigma(T)$ produces the spectral decomposition

(16.6) $$X = X(T,\overline{\Delta}_0) + \sum_{j=1}^{m} \Xi(T,\overline{\Delta}_j).$$

Since, for $1 \leq j \leq m$, $\overline{\Delta}_j$ is compact, (III) and 13.7 imply

(16.7) $$\Xi(T,\overline{\Delta}_j) \subset E(G), \quad 1 \leq j \leq m.$$

Let $x \in E(G)$ be given. In view of (16.6), there is a representation

(16.8) $$x = \sum_{j=0}^{m} x_j \quad \text{with} \quad x_0 \in X(T,\overline{\Delta}_0); \quad x_j \in \Xi(T,\overline{\Delta}_j), \quad 1 \leq j \leq m.$$

It follows from (16.8) and (16.7) that $x_0 \in E(G) \cap X(T,\overline{\Delta}_0)$ and hence the decomposition (16.5) is obtained.

Part C. Since $E(G)$ and $X(T,\overline{\Delta}_0)$ are analytically invariant under T (Lemma 16.4), we obtain successively

$$\sigma[T|E(G) \cap X(T,\overline{\Delta}_0)] \subset \overline{G} \cap \overline{\Delta}_0 \subset \overline{G} \cap [\overline{H}_0 \cup (\bigcup_{j=0}^{m} \overline{\delta}_j) \cup G^c]$$

$$= \overline{G} \cap [\overline{H}_0 \cup (\bigcup_{j=0}^{m} \overline{\delta}_j)] \subset \overline{H}_0 \cup (\bigcup_{j=0}^{m} \overline{\delta}_j).$$

Since, by construction (I), $a_0 < a_2$ and $b_{m-1} < b_0$, the sets $\overline{\delta}_j$ $(2 \leq j \leq m-1)$ do not intersect $\overline{H}_0$. Thus, the sets $(\overline{H}_0 \cup \overline{\delta}_1 \cup \overline{\delta}_m)$, $\overline{\delta}_2, \ldots \overline{\delta}_{m-1}$ are mutually disjoint. There correspond subspaces $Y_0, Y_2, \ldots, Y_{m-1}$, invariant under $T|E(G) \cap X(T,\overline{\Delta}_0)$, producing the direct sum

(16.9) $$E(G) \cap X(T,\overline{\Delta}_0) = Y_0 \oplus Y_2 \oplus \ldots \oplus Y_{m-1};$$

with

(16.10) $$\sigma(T|Y_0) \subset \overline{H}_0 \cup \overline{\delta}_1 \cup \overline{\delta}_m;$$

(16.11) $$\sigma(T|Y_j) \subset \overline{\delta}_j, \quad 2 \leq j \leq m-1.$$

Since

$$\overline{H}_0 \cup \overline{\delta}_1 \cup \overline{\delta}_m \cup \overline{\Delta}_1 \cup \overline{\Delta}_m \subset G_0,$$

(16.10) and 12.6 imply

$$Y_0 \subset X(T,\overline{H}_0 \cup \overline{\delta}_1 \cup \overline{\delta}_m) \subset \overline{X(T,G_0)} \subset E(G_0),$$

$$\Xi(T,\bar{\Delta}_1) \subset \overline{X(T,G_0)} \subset E(G_0),$$
$$\Xi(T,\bar{\Delta}_m) \subset \overline{X(T,G_0)} \subset E(G_0).$$

Thus, by (16.7) and (16.9), we have

(16.12) $$Y_0 + \Xi(T,\bar{\Delta}_1) + \Xi(T,\bar{\Delta}_m) \subset E(G) \cap E(G_0).$$

Similarly, since

$$\bar{\delta}_j \cup \bar{\Delta}_j \subset \bar{H}_j \subset G_i \quad \text{for } 2 \le j \le m-1 \text{ and some } i \;\; (0 \le i \le n),$$

12.6, (16.9) and (16.11) imply

(16.13) $$Y_j + \Xi(T,\bar{\Delta}_j) \subset E(G) \cap E(G_i) \quad \text{for } 2 \le j \le m-1 \text{ and}$$

some i $(0 \le i \le n)$.

Now, it follows from (16.5), (16.9), (16.12) and (16.13) that

$$E(G) = \sum_{i=0}^{n} E(G) \cap E(G_i). \quad \square$$

For bounded operators, the foregoing theorem gives rise to a more powerful property.

16.6. COROLLARY. Let $T \in B(X)$ be decomposable with $\sigma(T) \subset \mathbb{R}$. If, for $Y \in \operatorname{Inv} T$, there exists $G_0 \in G^{\mathbb{R}}$ such that

(16.14) $$\overline{X(T,G_0)} \subset Y \subset X(T,\bar{G}_0),$$

then $T|Y$ is decomposable. In particular, $T|\overline{X(T,G_0)}$ and $T|X(T,\bar{G}_0)$ are decomposable.

PROOF. For $Y(\in \operatorname{Inv} T) = E(G)$, the proof of 16.4 holds and hence Y is analytically invariant under T. Assuming that there is $G_0 \in G^{\mathbb{R}}$ such that (16.14) holds, Y is analytically invariant under $T|X(T,\bar{G}_0)$. In particular, Y is a ν-space of $T|X(T,\bar{G}_0)$ and hence

(16.15) $$\sigma(T|Y) \subset \sigma[T|X(T,\bar{G}_0)] \subset \bar{G}_0.$$

For $G \in G^{\mathbb{R}}$, put

(16.16) $$E(G) = \begin{cases} Y, & \text{if } G = G_0; \\ X(T,\bar{G}), & \text{if } G \ne G_0. \end{cases}$$

Then, since T is decomposable, (16.15) and (16.16) imply that E is a spectral resolvent of T. Thus, by 16.5, E is a strong spectral resolvent and hence $T|Y = T|E(G_0)$ is decomposable. $\square$

§.17. SPECTRAL DECOMPOSITIONS WITH RESPECT TO THE IDENTITY

The conceptual roots of the contemporary spectral theory lie in the resolution of the identity, as formalized and developed by Dunford. With possible applications in mind, one can build a strong case for the urgency of generalizing Dunford's approach to the spectral decomposition problem on the grounds that it promises to be useful and mathematically appealing.

Given a closed operator T, we shall denote by CM(T) the collection of all operators in B(X) which commute with T and we shall use the notation H Inv T for the lattice of all subspaces of X which are hyperinvariant under T.

17.1. DEFINITION. Given T and $n \in \mathbb{N}$, we say that T has the n-*spectral decomposition property with respect to the identity* (n-SDI) if, for each $\{G_i\}_{i=0}^n \in \operatorname{cov} \sigma(T)$ with $G_0 \in V_\infty$ and $\{G_i\}_{i=1}^n \subset G^K$, there exists a system $\{X_i\}_{i=0}^n \subset \operatorname{Inv} T$ satisfying the following conditions:

(I) $\sigma(T|X_i) \subset G_i$, $0 \leq i \leq n$;

(II) $X_i \subset \mathcal{D}_T$, if $G_i \in G^K$ $(1 \leq i \leq n)$;

(III) there exists $\{P_i\}_{i=0}^n \subset CM(T)$ such that

$$I = \sum_{i=0}^{n} P_i \quad \text{and} \quad R(P_i) \subset X_i, \quad 0 \leq i \leq n.$$

If, for every $n \in \mathbb{N}$, T has the n-SDI then we say that T has the *spectral decomposition property with respect to the identity* (SDI).

Clearly, the n-SDI (SDI) implies the n-SDP (SDP). Thus, the properties expressed by 5.9 and 4.34 are carried over to operators with the n-SDI, as follows.

Given T with the n-SDI, for every closed F, X(T,F) is closed and, for F compact, the direct sum decomposition holds

$$X(T,F) = \Xi(T,F) \oplus X(T,\emptyset),$$

with

$$\sigma[T|\Xi(T,F)] = \sigma[T|X(T,F)].$$

Consequently, the invariant subspaces X_i in 17.1 can be replaced by $X_0 = X(T,\overline{G}_0)$ and $X_i = \Xi(T,\overline{G}_i)$, $1\le i\le n$.

The equivalence of the 1-SDI and the SDI follows from

17.2. THEOREM. If T has the n-SDI then T has the (n+1)-SDI.

PROOF. Let $\{G_i\}_{i=0}^{n+1}$ e cov $\sigma(T)$ with G_0 e V_∞ and $\{G_i\}_{i=1}^{n+1} \subset G^K$. Let H_0 e V_∞ be such that $\overline{H}_0 \subset G_0$, $\{H_0,G_i\}_{i=1}^{n+1}$ e cov $\sigma(T)$. A set H_1 e G^K is chosen such that $\{G_0,H_1\}$ e cov $\sigma(T)$, $\{G_i\}_{i=1}^{n+1}$ e cov $\overline{H}_1$ and $\overline{H}_0 \cap \overline{H}_1 = \emptyset$. Assuming that T has the n-SDI, it also has the 1-SDI and hence there exist P,Q e CM(T) satisfying the following properties:

$$I = P + Q;$$

$$R(P) \subset X(T,\overline{G}_0),\quad R(Q) \subset \Xi(T,\overline{H}_1).$$

Since T has the n-SDI and $\{H_0 \cup G_{n+1},G_i\}_{i=1}^{n}$ e cov $\sigma(T)$, there exists $\{Q_i\}_{i=0}^{n} \subset CM(T)$ such that

$$I = \sum_{i=0}^{n} Q_i;$$

$$R(Q_0) \subset X(T,\overline{H_0 \cup G_{n+1}});\quad R(Q_i) \subset \Xi(T,\overline{G}_i),\quad 1\le i\le n.$$

Moreover, the subspaces $X(T,\overline{H_0 \cup G_{n+1}})$, $\Xi(T,\overline{G}_i)$ $(1\le i\le n)$ being hyperinvariant under T, they are invariant under P, Q, Q_i $(0\le i\le n)$. Put $P_0 = P$, $P_i = QQ_i$ $(1\le i\le n)$, $P_{n+1} = QQ_0$ and obtain

$$(17.1)\quad \begin{cases} R(P_0) = R(P) \subset X(T,\overline{G}_0); \\ R(P_i) = R(QQ_i) \subset \Xi(T,\overline{H}_1) \cap \Xi(T,\overline{G}_i) \subset \Xi(T,\overline{G}_i),\quad 1\le i\le n; \\ R(P_{n+1}) = R(QQ_0) \subset \Xi(T,\overline{H}_1) \cap X(T,\overline{H_0 \cup G_{n+1}}) \\ \subset \Xi[T,\overline{H}_1 \cap (\overline{H}_0 \cup \overline{G}_{n+1})] \subset \Xi(T,\overline{G}_{n+1}); \end{cases}$$

$$(17.2)\quad \sum_{i=0}^{n+1} P_i = P + Q(\sum_{i=0}^{n} Q_i) = P + Q = I.$$

Thus, it follows from (17.1) and (17.2) that T has the (n+1)-SDI. □

Following the analogy with the SDP, the open cover $\{G_i\}_{i=0}^n$, as given in 17.1, can be taken as an element of $\operatorname{cov} \mathbb{C}$ instead of an element of $\operatorname{cov} \sigma(T)$.

17.3. THEOREM. Given T, the following assertions are equivalent:

(I) T has the SDI;

(II) (a) T has the SDP;

(b) for every closed F and each open $G(\supset F) \in V_\infty$, there exists $P \in CM(T)$ such that

$$Px = x \quad \text{for all} \quad x \in X(T,F) \quad \text{and} \quad R(P) \subset X(T,\overline{G});$$

(III) (a) T has the SDP;

(b) for every closed F and each open $G(\supset F) \in V_\infty$, there exists $R_\lambda \in CM(T)$, analytic for $\lambda \notin \overline{G}$, with the following properties

$$(\lambda-T)R_\lambda x = x \quad \text{for all} \quad x \in X(T,F) \quad \text{and} \quad R(R_\lambda) \subset X(T,\overline{G}) \cap \mathcal{D}_T.$$

PROOF. (I) => (II): Given $G(\supset F) \in V_\infty$, choose $G_1 \in G^K$ such that $\{G,G_1\} \in \operatorname{cov} \sigma(T)$ and $\overline{G}_1 \cap F = \emptyset$. There exist $P,P_1 \in CM(T)$ such that

$$I = P + P_1, \quad R(P) \subset X(T,\overline{G}), \quad R(P_1) \subset \Xi(T,\overline{G}_1).$$

Let $x \in X(T,F)$. We have

$$P_1x \in \Xi(T,\overline{G}_1) \cap X(T,F) = \Xi(T,\overline{G}_1 \cap F) = \Xi(T,\emptyset) = \{0\}$$

and hence $P_1x = 0$. This implies that $x = Px + P_1x = Px$. Thus, P satisfies the asserted conditions of (II,b).

(II) => (III): Let P be the operator given in (II,b). Then

$$R_\lambda = R[\lambda;T|X(T,\overline{G})]P \quad \text{for} \quad \lambda \notin \overline{G}$$

satisfies conditions (III,b).

(III) => (I): Let $\{G_0,G_1\} \in \operatorname{cov} \sigma(T)$ with $G_0 \in V_\infty$ and $G_1 \in G^K$. Choose an open $H_0 \in V_\infty$ with the properties:

$$\overline{H}_0 \subset G_0 \quad \text{and} \quad \{H_0,G_1\} \in \operatorname{cov} \sigma(T).$$

By hypothesis, there is $R_\lambda \in CM(T)$, $\lambda \notin \overline{G}_0$ satisfying conditions

(17.3) $$(\lambda-T)R_\lambda x = x \quad \text{for} \quad x \in X(T,\overline{H}_0);$$

(17.4) $$R(R_\lambda) \subset X(T,\overline{G}_0) \cap \mathcal{D}_T.$$

Fix $\lambda \notin \overline{G}_0$, put $P_0 = (\lambda-T)R_\lambda$, $P_1 = I - P_0$ and note that $P_0, P_1 \in CM(T)$. It follows from (17.3) that

$$P_0 x = x \quad \text{for } x \in X(T,\overline{H}_0) \tag{17.5}$$

and (17.4) implies that $R(P_0) \subset X(T,\overline{G}_0)$. It remains to show that

$$R(P_1) \subset \Xi(T,\overline{G}_1). \tag{17.6}$$

Since, by hypothesis, T has the SDP,

$$X = X(T,\overline{H}_0) + \Xi(T,\overline{G}_1)$$

implies that each $x \in X$ has a representation

$$x = x_0 + x_1, \quad x_0 \in X(T,\overline{H}_0), \quad x_1 \in \Xi(T,\overline{G}_1).$$

In view of (17.5), $P_0 x_0 = x_0$ and hence $P_1 x_0 = 0$. Therefore,

$$P_1 x = P_1 x_0 + P_1 x_1 = P_1 x_1 \in \Xi(T,\overline{G}_1),$$

which leads one to inclusion (17.6).

Thus, for $X_0 = X(T,\overline{G}_0)$, $X_1 = \Xi(T,\overline{G}_1)$ and P_0, P_1, as defined above, all conditions asserted in 17.1 are satisfied. □

17.4. REMARK. If F is compact in conditions (II,b) and (III,b) of the foregoing theorem, then they can be replaced by the following:

(II') (b) For every compact F and $G(\supset F) \in G^K$, there exists $P \in CM(T)$ such that

$$Px = x \quad \text{for all } x \in \Xi(T,F) \text{ and } R(P) \subset \Xi(T,\overline{G});$$

(III') (b) For every compact F and $G(\supset F) \in G^K$, there exists $R_\lambda \in CM(T)$, $\lambda \notin \overline{G}$ such that

$$(\lambda-T)R_\lambda x = x \quad \text{for all } x \in \Xi(T,F) \text{ and } R(R_\lambda) \subset \Xi(T,\overline{G}).$$

17.5. THEOREM. Given T with the SDI, let $Z \in H\ \mathrm{Inv}\ T$. Then $\hat{T} = T/Z$ is closed on X/Z.

PROOF. We present the proof in three parts.

Part A. Let $F(\neq \mathbb{C})$ be a closed neighborhood of ∞ and let $G \supset F$ be open with $\overline{G} \neq \mathbb{C}$. By 17.3, there exists $P \in CM(T)$ such that

$$Px = x \quad \text{for all } x \in X(T,F) \text{ and } R(P) \subset X(T,\overline{G}).$$

Put $R_\lambda = R[\lambda;T|X(T,\overline{G})]P$, $\lambda \notin \overline{G}$. Then

$$R(R_\lambda) \subset X(T,\overline{G}) \cap \mathcal{D}_T \tag{17.7}$$

and, for every $x \in X$, we have

$$(\lambda-T)R_\lambda x = Px.$$

Then $TR_\lambda \in B(X)$ and this implies that $\hat{T}\hat{R}_\lambda = (TR_\lambda)^\wedge \in B(X/Z)$. Furthermore, for each $\hat{x} \in X/Z$,

(17.8) $$(\lambda-\hat{T})\hat{R}_\lambda\hat{x} = \hat{P}\hat{x}.$$

It follows from (17.7), that $R(PR_\lambda) \subset X(T,\overline{G}) \cap \mathcal{D}_T$ and hence

(17.9) $$PR_\lambda = R[\lambda;T|X(T,\overline{G})](\lambda-T)PR_\lambda = R[\lambda;T|X(T,\overline{G})]P(\lambda-T)R_\lambda = R_\lambda P.$$

Put

(17.10) $$\hat{X}_{\overline{G}} = \{\hat{x} \in X/Z : \hat{P}\hat{x} = \hat{x}\}$$

and obtain

(17.11) $$\hat{X}_{\overline{G}} \supset [X(T,F)]^\wedge = \{\hat{x} \in X/Z : \hat{x} \cap X(T,F) \neq \emptyset\}.$$

Since $\hat{P}$ commutes with $\hat{T}$,

$$\hat{P}\hat{T}\hat{x} = \hat{T}\hat{P}\hat{x} = \hat{T}\hat{x} \quad \text{for all } \hat{x} \in \hat{X}_{\overline{G}} \cap \mathcal{D}_{\hat{T}}.$$

Consequently, $\hat{X}_{\overline{G}}$ is invariant under $\hat{T}$. It follows from (17.9) that $\hat{P}$ commutes with $\hat{R}_\lambda$ and hence $\hat{X}_{\overline{G}}$ is invariant under $\hat{R}_\lambda$. Since $\hat{R}_\lambda \in CM(\hat{T})$, (17.8) implies

(17.12) $$\begin{cases} (\lambda-\hat{T})\hat{R}_\lambda\hat{x} = \hat{x} & \text{for } \hat{x} \in \hat{X}_{\overline{G}}; \\ \hat{R}_\lambda(\lambda-\hat{T})\hat{x} = \hat{x} & \text{for } \hat{x} \in \hat{X}_{\overline{G}} \cap \mathcal{D}_{\hat{T}}. \end{cases}$$

Thus, we have

$$R(\lambda;\hat{T}|\hat{X}_{\overline{G}}) = \hat{R}_\lambda|\hat{X}_{\overline{G}}\,.$$

Since $\hat{R}_\lambda|\hat{X}_{\overline{G}} \in B(\hat{X}_{\overline{G}})$, $\lambda-\hat{T}|\hat{X}_{\overline{G}}$ is closed and so is $\hat{T}|\hat{X}_{\overline{G}}$. Moreover, (17.12) implies that $\sigma(\hat{T}|\hat{X}_{\overline{G}}) \subset \overline{G}$.

Part B. Let F be compact and let $G \supset F$ be open relatively compact. It follows from 17.4 (II') (b) that there exists $P \in CM(T)$ such that

$$Px = x \quad \text{for all } x \in \Xi(T,F) \quad \text{and} \quad R(P) \subset \Xi(T,\overline{G}).$$

Put

(17.13) $$\hat{\Xi}_{\overline{G}} = \{\hat{x} \in X/Z : \hat{P}\hat{x} = \hat{x}\}$$

and obtain

(17.14) $$\hat{\Xi}_{\overline{G}} \supset [\Xi(T,F)]^\wedge = \{\hat{x} \in X/Z : \hat{x} \cap \Xi(T,F) \neq \emptyset\}.$$

With a technique, similar to that used in Part A, one can show that $\hat{\Xi}_{\overline{G}} \in \text{Inv}\,\hat{T}$, $\hat{T}|\hat{\Xi}_{\overline{G}}$ is closed and $\sigma(\hat{T}|\hat{\Xi}_{\overline{G}}) \subset \overline{G}$. It follows from

$R(P) \subset \Xi(T,\overline{G}) \subset \mathcal{D}_T$ that $R(\hat{P}) \subset \mathcal{D}_{\hat{T}}$ and hence $\hat{\Xi}_{\overline{G}} \subset \mathcal{D}_{\hat{T}}$, by (17.13). Therefore $\hat{T}|\hat{\Xi}_{\overline{G}}$ is bounded.

Part C. In this final part of the proof, we show that $\hat{T}$ is closed. Let $\{G,H\} \in \operatorname{cov} \sigma(T)$ with $G \in V_\infty$, $\overline{G} \neq \mathbb{C}$ and $H \in G^K$. Let $\{G',H'\} \in \operatorname{cov} \sigma(T)$ with $G' \in V_\infty$, $\overline{G'} \subset G$ and $\overline{H'} \subset H$. We have

$$X = X(T,\overline{G'}) + \Xi(T,\overline{H'}).$$

On the quotient space X/Z, there corresponds

$$\hat{X} = [X(T,\overline{G'})]^\wedge + [\Xi(T,\overline{H'})]^\wedge. \tag{17.15}$$

Apply (17.11) and (17.14) to the pairs $\overline{G'}$, G and $\overline{H'}$, H, respectively, to write

$$[X(T,\overline{G'})]^\wedge \subset \hat{X}_{\overline{G}}, \quad [\Xi(T,\overline{H'})]^\wedge \subset \hat{X}_{\overline{H}}$$

and deduce from (17.15) that

$$\hat{X} = \hat{X}_{\overline{G}} + \hat{\Xi}_{\overline{H}}.$$

$\hat{T}|\hat{X}_{\overline{G}}$ is closed by Part A and $\hat{T}|\hat{\Xi}_{\overline{H}}$ is bounded by Part B of the proof. Thus $\hat{T}$ is closed by 5.15. □

17.6. THEOREM. Given T with the SDI, let $Y,Z \in H \operatorname{Inv} T$ with $Y \supset Z$. Then $\hat{T}|\hat{Y} = (T/Y)|(Y/Z)$ has the SDI.

PROOF. First, we show that, for open $G \in V_\infty$ with $\overline{G} \neq \mathbb{C}$,

$$\sigma(\hat{T}|\hat{Y} \cap \hat{X}_{\overline{G}}) \subset \overline{G} \tag{17.16}$$

and, for $G \in G^K$,

$$\sigma(\hat{T}|\hat{Y} \cap \hat{\Xi}_{\overline{G}}) \subset \overline{G}, \tag{17.17}$$

where $\hat{X}_{\overline{G}}$ and $\hat{\Xi}_{\overline{G}}$ are defined by (17.10) and (17.13), respectively. We confine this segment of the proof to (17.16).

Since $Y \in H \operatorname{Inv} T$, we have $Y \in \operatorname{Inv} R_\lambda$, where for $\lambda \notin \overline{G}$, $R_\lambda = R[\lambda;T|X(T,\overline{G})]P$. It follows that $\hat{Y} \in \operatorname{Inv} \hat{R}_\lambda$. Since $\hat{R}_\lambda \in CM(\hat{T})$, by (17.8), we have

$$(\lambda-\hat{T})\hat{R}_\lambda\hat{x} = \hat{x} \quad \text{for } \hat{x} \in \hat{Y} \cap \hat{X}_{\overline{G}};$$

$$\hat{R}_\lambda(\lambda-\hat{T})\hat{x} = \hat{x} \quad \text{for } \hat{x} \in \hat{Y} \cap \hat{X}_{\overline{G}} \cap \mathcal{D}_{\hat{T}}$$

and hence (17.16) follows.

Next, assume that $\{G,H\} \in \operatorname{cov} \mathbb{C}$ with $G \in V_\infty$, $\overline{G} \neq \mathbb{C}$ and $H \in G^K$. Let $\{G',H'\}$ cov $\mathbb{C}$ with $G' \in V_\infty$, $\overline{G'} \subset G$, $\overline{H'} \subset H$. By the SDI of

T, there exist $P_G, P_H \in CM(T)$ with

$$I = P_G + P_H, \quad R(P_G) \subset X(T,\overline{G}'), \quad R(P_H) \subset \Xi(T,\overline{H}').$$

On the quotient space X/Z, there correspond

$$(17.18) \qquad \hat{I} = \hat{P}_G + \hat{P}_H, \quad R(\hat{P}_G) \subset [X(T,\overline{G}')]^\wedge \subset \hat{X}_{\overline{G}}, \quad R(\hat{P}_H) \subset [\Xi(T,\overline{H}')]^\wedge \subset \hat{\Xi}_{\overline{H}}.$$

By restricting the operators in (17.18) to $\hat{Y}$, we obtain

$$(17.19) \qquad \hat{I}|\hat{Y} = \hat{P}_G|\hat{Y} + \hat{P}_H|\hat{Y}, \quad R(\hat{P}_G|\hat{Y}) \subset \hat{Y} \cap \hat{X}_{\overline{G}}, \quad R(\hat{P}_H|\hat{Y}) \subset \hat{Y} \cap \hat{\Xi}_{\overline{H}}.$$

Now (17.19), (17.16) and (17.17) imply that $\hat{T}|\hat{Y}$ has the SDI. □

17.7. COROLLARY. If T has the SDI then every $Y \in H\ Inv\ T$ is analytically invariant under T.

PROOF. $\hat{T}$ has the SDI, by 17.6, hence it has the SVEP. Thus $Y \in AI(T)$, by 4.14. □

17.8. COROLLARY. It T has the SDI then, for any $Y \in H\ Inv\ T$, $T|Y$ has the SDI.

PROOF follows directly from 17.6, for $Z = \{0\}$. □

17.9. COROLLARY. If T has the SDI then, for any pair $Y,Z \in H\ Inv\ T$ with $Y \supset Z$, we have $\sigma(T|Y) \supset \sigma(T|Z)$.

PROOF. $Z \in AI(T)$, by 17.7 and hence $Z \in AI(T|Y)$. In particular, Z is a ν-space of $T|Y$ and the asserted inclusion follows. □

17.10. COROLLARY. If T has the SDI then T has the SSDP.

PROOF. Since every spectral maximal space of T is hyperinvariant under T (Proposition 4.23), the assertion of the corollary follows from 17.8. □

17.11. PROPOSITION. Let T be densely defined and have the SDP. If, for every $G \in G^K$, there exists $P \in CM(T)$ such that

$$(17.20) \qquad R(P) \subset \Xi(T,\overline{G}).$$

Then $P^* \in CM(T^*)$ and

$$(17.21) \qquad R(P^*) \subset \Xi^*(T^*,\overline{G}).$$

PROOF. Since T has the SDP, it follows from 9.8 that, for every $G \in G^K$,

$$(17.22) \qquad \Xi^*(T^*,\overline{G}) = X(T,H)^a, \quad \text{where } H = (\overline{G})^c.$$

By hypothesis, for $G \in G^K$, there is $P \in CM(T)$ satisfying (17.20). For $x \in X(T,H)$, $\sigma(x,T) \cap \overline{G} = \emptyset$. It follows from $\sigma(Px,T) \subset \sigma(x,T)$ that

$$Px \in \Xi(T,\overline{G}) \cap X[T,\sigma(x,T)] = \Xi[T,\overline{G} \cap \sigma(x,T)] = \Xi(T,\emptyset) = \{0\}$$

and hence $Px = 0$. Let $x^* \in X^*$. Then, for $x \in X(T,H)$, we have

$$\langle x, P^*x^* \rangle = \langle Px, x^* \rangle = 0.$$

Thus (17.22) implies that $P^*x^* \in \Xi^*(T^*,\overline{G})$ and hence inclusion (17.21) holds. □

17.12. COROLLARY. If T is densely defined and has the SDI, then T* has the SDI.

PROOF. Let $\{G_0,G_1\} \in \operatorname{cov} \mathbb{C}$ with $G_0 \in V_\infty$ and $G_1 \in G^K$. There are $P_i \in CM(T)$, $i=0,1$ such that

$$I = P_0 + P_1, \quad R(P_0) \subset X(T,\overline{G}_0), \quad R(P_1) \subset \Xi(T,\overline{G}_1).$$

Then

$$I^* = P_0^* + P_1^* \tag{17.23}$$

and 17.11 implies

$$R(P_1^*) \subset \Xi^*(T^*,\overline{G}_1). \tag{17.24}$$

Using an argument similar to that in the proof of 17.11, we obtain

$$R(P_0^*) \subset X^*(T^*,\overline{G}_0). \tag{17.25}$$

In view of (17.23), (17.24) and (17.25), T* has the SDI. □

17.13. DEFINITION. Given T with the SDP. If there exist $\{G_n\}_{n=1}^\infty \subset G^K$ and $\{P_n\}_{n=1}^\infty \subset CM(T)$ with

$$R(P_n) \subset \Xi(T,\overline{G}_n), \quad n \in \mathbb{N} \quad \text{and} \quad \lim_{n\to\infty} \langle P_n x, x^* \rangle = \langle x, x^* \rangle,$$

for every $x \in X$ and $x^* \in X^*$, then we say that T has *property* (δ).

17.14. THEOREM. Given T with the SDP. If T has property (δ) then, for every family $\{X_\alpha\}_{\alpha\in A} \subset H \operatorname{Inv} T$, we have $Y = \bigvee_{\alpha\in A} X \in H \operatorname{Inv} T$.

PROOF. Since every $S \in CM(T)$ is bounded, Y is invariant under S. Therefore, it suffices to show that Y is invariant under T.

Let $x \in X_\alpha$. Then $P_n x \in X_\alpha \cap \Xi(T,\overline{G}_n)$ and

$$\lim_{n\to\infty} \langle P_n x, x^* \rangle = \langle x, x^* \rangle. \tag{17.26}$$

By (17.26) and the Hahn-Banach theorem,

$$X_\alpha = \bigvee_{n=1}^{\infty} [X_\alpha \cap \Xi(T,\overline{G}_n)].$$

Consequently, $X_\alpha \cap \Xi(T,\overline{G}_n) \subset \mathcal{D}_T$ implies that $T|X_\alpha$ is densely defined. In particular, T is densely defined. For $x \in X_\alpha \cap \mathcal{D}_T$ and $x^* \in (X_\alpha)^a \cap \mathcal{D}_{T^*}$, the equalities

$$\langle x,T^*x^*\rangle = \langle Tx,x^*\rangle = 0 \text{ and } \overline{X_\alpha \cap \mathcal{D}_T} = X_\alpha$$

imply that $T^*x^* \in (X_\alpha)^a$ and hence

(17.27) $$(X_\alpha)^a \in \text{Inv } T^*.$$

Since $Y^a = \bigcap_{\alpha \in A} (X_\alpha)^a$, it follows from (17.27) that $Y^a \in \text{Inv } T^*$. For $x^* \in Y^a$, 17.11 implies that $P_n^* x^* \in \Xi^*(T^*,\overline{G}_n)$ for all n, and hence we have

$$P_n^*x^* \in Y^a \cap \Xi^*(T^*,\overline{G}_n) \subset Y^a \cap \mathcal{D}_{T^*}, \quad T^*P_n^*x^* \in Y^a, \quad n \in \mathbb{N}.$$

Thus, for every n and each $x \in Y \cap \mathcal{D}_T$, we have

$$0 = \langle x,T^*P_n^*x^*\rangle = \langle Tx,P_n^*x^*\rangle = \langle P_nTx,x^*\rangle .$$

Consequently,

$$\langle Tx,x^*\rangle = \lim_{n\to\infty} \langle P_nTx,x^*\rangle = 0$$

and this implies that Y is invariant under T. □

We continue our study of operators with the SDI, in terms of maximal nets of hyperinvariant subspaces.

17.15. DEFINITION. Given T and a totally ordered set A, a family $N = \{X_\alpha\}_{\alpha \in A} \subset H \text{ Inv } T$ is called a *net of hyperinvariant subspaces* under T if, for $\alpha,\beta \in A$, $\alpha < \beta$ implies $X_\alpha \subsetneq X_\beta$.

By Zorn's lemma, there is a maximal net of hyperinvariant subspaces under T which contains N. Without loss of generality, we may assume that N itself is the maximal net. Since $\{\{0\},X\} \subset N$, N is non-empty.

17.16. LEMMA. Let T have the SDP and property (δ). Let $N = \{X_\alpha\}_{\alpha \in A}$ be a maximal net of hyperinvariant subspaces under T. Then, for every $\alpha \in A$, there exists $\beta \in A$ such that $X_\beta = \bigvee_{\gamma<\alpha} X_\gamma$.

PROOF. Put $Y = \bigvee_{\gamma<\alpha} X_\gamma$. Then $Y \in H \text{ Inv } T$, by 17.14, and

(17.28) $$X_\gamma \subset Y \subset X_{\gamma'}, \quad \text{for } \gamma < \alpha \le \gamma' .$$

Suppose to the contrary that $Y \notin N$. Then, in view of (17.28), the net $M = \{X_\alpha, Y\}_{\alpha \in A} \subset H \text{ Inv } T$ properly contains N. This, however, contradicts the maximality of N. Thus, there is $\beta \in A$ such that $Y = X_\beta$. □

Denoting $\beta = \alpha - 0 \leq \alpha$, we say that N is continuous at α if $\alpha - 0 = \alpha$, otherwise, we say that N is discontinuous at α.

17.17. THEOREM. Let T have the SDI and be endowed with property (δ). Let $N = \{X_\alpha\}_{\alpha \in A}$ be a maximal net of hyperinvariant subspaces under T. Then

(i) for every $\alpha \in A$, we have

$$\sigma(T|X_\gamma) \subset \sigma(T|X_{\alpha-0}) \quad \text{for} \quad \gamma < \alpha; \tag{17.29}$$

$$\sigma(T|X_{\alpha-0}) = \bigvee\{\sigma(T|X_\gamma) : \gamma < \alpha\}; \tag{17.30}$$

$$\sigma(T|X_{\alpha-0}) \subset \sigma(T|X_\alpha); \tag{17.31}$$

(ii) if N is discontinuous at α, let $\hat{T}_\alpha$ be the coinduced operator by $T|X_\alpha$ on the quotient space $\hat{X}_\alpha = X_\alpha / X_{\alpha-0}$. Then

either (a) $\hat{T}_\alpha$ is unbounded and $\sigma(\hat{T}_\alpha) = \emptyset$,

or (b) $\hat{T}_\alpha \in B(X_\alpha)$ and $\sigma(\hat{T}_\alpha)$ consists of exactly one point ξ_α. Moreover, either $\hat{T}_\alpha = \xi_\alpha \hat{I}_\alpha$ or $\hat{T}_\alpha - \xi_\alpha \hat{I}_\alpha$ is quasinilpotent.

(iii) N is discontinuous at α if

either (a) $\sigma(T|X_{\alpha-0}) \neq \sigma(T|X_\alpha)$,

or (b) $\hat{X}_\alpha$ is finite dimensional.

In case (a) there exists $Y_\alpha \in \text{Inv } T$ such that

$$X_\alpha = X_{\alpha-0} \oplus Y_\alpha \tag{17.32}$$

and $\sigma(T|Y_\alpha) = \{\xi_\alpha\}$. In case (b), $\hat{X}_\alpha$ being finite dimensional, we have $Y_\alpha \subset X_\alpha$ such that (17.32) remains valid.

If $P_\alpha \in B(X_\alpha)$ is the projection onto Y_α along $X_{\alpha-0}$ then, for every $x \in Y_\alpha$, we have $P_\alpha Tx = \xi_\alpha x$.

PROOF. (i): Since T has the SDI, inclusions (17.29) and (17.31) evidently hold. We always have

$$\sigma(T|X_{\alpha-0}) \supset \bigvee\{\sigma(T|X_\gamma) : \gamma < \alpha\}. \tag{17.33}$$

Suppose to the contrary that the opposite of (17.33) is not true. Then, there exists $\xi \in \sigma(T|X_{\alpha-0}) - \bigvee_{\gamma<\alpha} \sigma(T|X_\gamma)$. Let $\{G_0, G_1\} \in \text{cov } \sigma(T|X_{\alpha-0})$

with $G_0 \in V_\infty$ and $G_1 \in G^K$, be such that

$$G_0 \supset \bigvee_{\gamma<\alpha} \sigma(T|X_\gamma), \quad \xi \notin \overline{G}_0 \quad \text{and} \quad \overline{G}_1 \cap \bigvee_{\gamma<\alpha} \sigma(T|X_\gamma) = \emptyset.$$

Since $S = T|X_{\alpha-0}$ has the SDI,

$$X_{\alpha-0} = X_{\alpha-0}(S,\overline{G}_0) + \Xi_{\alpha-0}(S,\overline{G}_1).$$

By 4.5,

$$\sigma(S) = \sigma[S|X_{\alpha-0}(S,\overline{G}_0)] \cup \sigma[S|\Xi_{\alpha-0}(S,\overline{G}_1)]$$

and hence

(17.34)
$$\sigma[S|X_{\alpha-0}(S,\overline{G}_0)] \supset \sigma(S) - \sigma[S|\Xi_{\alpha-0}(S,\overline{G}_1)]$$
$$\supset \sigma(T|X_{\alpha-0}) - \overline{G}_1 \supset \bigvee_{\gamma<\alpha} \sigma(T|X_\gamma).$$

Since $X_{\alpha-0}(S,\overline{G}_0)$ is a spectral maximal space of S, it follows from (17.34) that

$$X_\gamma \subset X_{\alpha-0}(S,\overline{G}_0) \quad \text{for} \quad \gamma < \alpha.$$

Thus, we have

$$X_{\alpha-0} = \bigvee_{\gamma<\alpha} X_\gamma \subset X_{\alpha-0}(S,\overline{G}_0).$$

Since, evidently, $X_{\alpha-0}(S,\overline{G}_0) \subsetneqq X_{\alpha-0}$, we ran into a contradiction and hence (17.30) holds.

(ii): $\hat{T}_\alpha$ has the SDI, by 17.6. First, assume that $\hat{T}_\alpha$ is unbounded. If $\sigma(\hat{T}_\alpha)$ has, at least, one point then there is a nontrivial $\hat{T}_\alpha$-bounded spectral maximal space $\hat{Z}$. Put

$$Z = \{x : x \in \hat{x}, \hat{x} \in \hat{Z}\}.$$

Then $X_{\alpha-0} \subsetneqq Z \subsetneqq X_\alpha$ and $Z \in$ H Inv T. This, however, is impossible because N is a maximal net. Thus, $\sigma(\hat{T}_\alpha) = \emptyset$.

Next, assume that $\hat{T}_\alpha$ is bounded and suppose to the contrary, that $\sigma(\hat{T}_\alpha)$ consists of more than one point. A similar argument to the one used above, leads one to a contradiction. Thus $\sigma(\hat{T}_\alpha)$ contains exactly one point ξ_α.

Further, suppose that the second statement of (ii) (b) is not true. Then, there is an integer $m > 1$ such that

$$\sigma(\hat{T}_\alpha - \xi_\alpha \hat{I}_\alpha)^m = \hat{0} \quad \text{and} \quad (\hat{T}_\alpha - \xi_\alpha \hat{I}_\alpha)^{m-1} \neq \hat{0}.$$

Letting

$$\hat{Z} = \overline{(\hat{T}_\alpha - \xi_\alpha \hat{I}_\alpha)^{m-1} \hat{X}_\alpha},$$

we have $Z = \{x : x \in \hat{x}, \hat{x} \in \hat{Z}\} \in H \text{ Inv } T$, with $X_{\alpha-0} \subsetneq Z \subsetneq X_\alpha$. But, this again contradicts the maximality of N. Thus, either $\hat{T}_\alpha = \xi_\alpha \hat{I}_\alpha$ or $\hat{T}_\alpha - \xi_\alpha \hat{I}_\alpha$ is quasinilpotent.

(iii): First, suppose that

$$\sigma(T|X_{\alpha-0}) \subsetneq \sigma(T|X_\alpha).$$

Then, $\sigma(\hat{T}_\alpha) \neq \emptyset$ by 3.1 and it consists of exactly one point ξ_α, with $\hat{T}_\alpha$ bounded, by (ii). Furthermore, we have

$$\sigma(T|X_\alpha) = \{\xi_\alpha\} \cup \sigma(T|X_{\alpha-0}), \quad \xi_\alpha \notin \sigma(T|X_{\alpha-0})$$

with $\{\xi_\alpha\}$ and $\sigma(T|X_{\alpha-0})$ spectral sets of $T|X_\alpha$. Consequently, the direct sum decomposition (17.32) follows by functional calculus.

Next, suppose that $\hat{X}_\alpha$ is finite dimensional. Then $\hat{T}_\alpha$ is bounded and $\sigma(\hat{T}_\alpha)$ consists of one point ξ_α. Let n be the dimension of $\hat{X}_\alpha$ and let $\{\hat{x}_1, \hat{x}_2, \dots, \hat{x}_n\}$ be a base of $\hat{X}_\alpha$. Then $\{x_1, x_2, \dots, x_n\}$ with $x_i \in \hat{x}_i$ $(1 \le i \le n)$ is a linearly independent system contained in X_α. The subspace Y_α, spanned by $\{x_1, x_2, \dots, x_n\}$, is topologically isomorphic to $\hat{X}_\alpha$ and (17.32) follows. By an argument, similar to that used in the proof of (ii,b), one obtains $\hat{T}_\alpha = \xi_\alpha \hat{I}_\alpha$. Since $P_\alpha T P_\alpha$ is similar to $\hat{T}_\alpha$, we have $P_\alpha Tx = \xi_\alpha x$, for $x \in Y_\alpha$. □

17.18. THEOREM. Given T, let $f : G \to \mathbb{C}$ be an analytic function on an open neighborhood G of $\sigma_\infty(T)$ satisfying the following conditions:

(a) f has, at most, two zeros: $f(0) = 0$, $f(\infty) = \lim_{\lambda \to \infty} f(\lambda) = 0$;

(b) f is nonconstant on every component of G;

(c) f(T) is a compact operator.

Then

(i) T has the SDI;

(ii) suppose that T has property (δ) and $N = \{X_\alpha\}_{\alpha \in A} \subset H \text{ Inv } T$ is a maximal net such that at every discontinuous point α, either $\sigma(\hat{T}_\alpha) = \{0\}$ or $\sigma(\hat{T}_\alpha) = \emptyset$, where $\hat{T}_\alpha$ is the coinduced operator by T_α on $X_\alpha / X_{\alpha-0}$. Then T is bounded and $\sigma(T) = \{0\}$, i.e. T is quasinilpotent.

PROOF. (i): It follows from the hypotheses on f that $\sigma(T)$ does not have a nonzero cluster point on $\mathbb{C}$. Then, as shown by Example 5.3, T decomposes X into a direct sum with the effect that T has the SDI.

(ii): First, we show that either $\sigma(T) = \{0\}$ or $\sigma(T) = \emptyset$.

Suppose to the contrary that there exists in $\mathbb{C}$ a point $\xi(\neq 0) \in \sigma(T)$. It follows from the hypotheses that ξ is an isolated point of $\sigma(T)$. Since both $\{0\}$ and X are in N and $\sigma(T|\{0\}) = \emptyset$, $\sigma(T|X) = \sigma(T)$, we divide the totally ordered set A into two parts:

(17.35) $$A^- = \{\gamma \in A : \xi \notin \sigma(T|X_\gamma)\};$$

(17.36) $$A^+ = \{\gamma \in A : \xi \in \sigma(T|X_\gamma)\}.$$

A^- and A^+ are nonempty and $A = A^- \cup A^+$, $A^- \cap A^+ = \emptyset$. Put

$$X^- = \bigvee\{X_\gamma : \gamma \in A^-\}, \quad X^+ = \bigcap\{X_\gamma : \gamma \in A^+\}.$$

The subspaces X^-, X^+ are hyperinvariant under T and they satisfy the inclusions

$$X_\gamma \subset X^- \subset X^+ \subset X_{\gamma'} \quad \text{for } \gamma \in A^-, \ \gamma' \in A^+.$$

Since N is a maximal net and T has property (δ), there exists $\alpha \in A$ such that

$$X^- = X_{\alpha-0}, \quad X^+ = X_\alpha.$$

Since ξ is an isolated point of $\sigma(T)$, it follows from (17.35) and (17.36) that

(17.37) $$\xi \notin \bigcup\{\sigma(T|X_\gamma) : \gamma \in A^-\} = \sigma(T|X_{\alpha-0}).$$

Next, we show that $\xi \in \sigma(T|X_\alpha)$. For every $\gamma \in A$, X_γ is analytically invariant under T, by 17.7. Thus it follows from 4.11, 4.9 and 4.34 that

(17.38) $$\Xi(T,\{\xi\}) \cap X_\gamma = \Xi_\gamma(T|X_\gamma,\{\xi\}), \qquad \gamma \in A.$$

Since $\xi \neq 0$ and $\xi \neq \infty$, $f(\xi) \neq 0$. Since $f(T)$ is a compact operator, $X[f(T),\{f(\xi)\}]$ is finite dimensional. Then 7.4 implies

$$X[f(T),\{f(\xi)\}] = \Xi[T,\{f^{-1}\circ f(\xi)\}] \supset \Xi(T,\{\xi\})$$

and hence $\Xi(T,\{\xi\})$ is finite dimensional. Since, for every $\gamma \in A^+$, we have $\xi \in \sigma(T|X_\gamma)$ and

$$\sigma[T|\Xi_\gamma(T|X_\gamma,\{\xi\})] = \sigma[T|X_\gamma(T|X_\gamma,\{\xi\})] = \{\xi\} \neq \emptyset,$$

it follows that

$$\Xi_\gamma(T|X_\gamma,\{\xi\}) \neq \{0\} \quad \text{for } \gamma \in A^+.$$

$\Xi(T,\{\xi\})$ being finite dimensional, (17.38) implies that there exists $\delta \in A^+$ such that, for every $\gamma \in A^+$ with $\gamma \leq \delta$, we have

(17.39) $$\Xi(T,\{\xi\}) \cap X_\gamma = \Xi(T,\{\xi\}) \cap X_\delta \neq \{0\}.$$

It follows from (17.38), (17.39) and from $X^+ = X_\alpha$ that

$$\Xi_\alpha(T|X_\alpha,\{\xi\}) = \Xi(T,\{\xi\}) \cap X_\alpha = \Xi(T,\{\xi\}) \cap (\cap\{X_\gamma : \gamma \in A^+\}) = \cap\{\Xi(T,\{\xi\}) \cap X_\gamma : \gamma \in A^+\} = \Xi(T,\{\xi\}) \cap X_\delta \neq \{0\}$$

and hence

$$\xi \in \sigma[T|\Xi_\alpha(T|X_\alpha,\{\xi\})] \subset \sigma(T|X_\alpha). \tag{17.40}$$

Relations (17.37) and (17.40) imply that N is discontinuous at α and $\sigma(\hat{T}_\alpha) = \{\xi\}$. By hypothesis, we have either $\sigma(\hat{T}_\alpha) = \{0\}$ or $\sigma(\hat{T}_\alpha) = \emptyset$. This contradicts the assumption that $\xi \neq 0$ and $\xi \neq \infty$. Therefore, $\sigma(T) = \{0\}$ or $\sigma(T) = \emptyset$. It follows from 4.34 that

$$X = X_0 \oplus X(T,\emptyset), \quad X_0 \subset \mathcal{D}_T$$

and $\sigma(T|X_0) = \{0\}$ or $\sigma(T|X_0) = \emptyset$ (the latter happens when $X_0 = \{0\}$). To conclude the proof, we have to show that $X(T,\emptyset) = \{0\}$. By property (δ), for every $n \in \mathbb{N}$, there exists $G_n \in G^K$ and $P_n \in CM(T)$ satisfying condition

$$R(P_n) \subset \Xi(T,\bar{G}_n).$$

For $x \in X(T,\emptyset)$, we have

$$P_n x \in \Xi(T,\bar{G}_n) \cap X(T,\emptyset) = \Xi(T,\emptyset) = \{0\}$$

and hence $P_n x = 0$. Then, it follows from

$$\langle x,x^*\rangle = \lim_{n\to\infty} \langle P_n x,x^*\rangle = 0, \quad x^* \in X^*,$$

that $x = 0$ and hence $X(T,\emptyset) = \{0\}$. Thus $\{0\} \neq X = X_0 \subset \mathcal{D}_T$ and

$$\sigma(T) = \sigma(T|X_0) = \{0\}. \ \square$$

In the spirit of 6.4, we extend the analogy between the SDP and SDI, by introducing the SDI-analogue of decomposable operator.

17.19. DEFINITION. If T has the SDI and $X(T,\emptyset) = \{0\}$, then we say that T is *decomposable with respect to the identity*.

Evidently, every bounded T with the SDI is decomposable with respect to the identity.

As an application, we shall consider a bounded operator T whose spectrum lies on a smooth Jordan curve. T has the SDP and hence T is decomposable. In particular, T has property (κ) and if F_1, F_2 are disjoint closed subsets of the plane, then

$$X(T,F_1 \cup F_2) = X(T,F_1) \oplus X(T,F_2)$$

by 5.12. The same conclusions are to be found in Radjabalipour's [R.1974] to which we owe some of the ideas used in the next theorem. Specifically, it was proved in [R.1974] that under the hypotheses of the following theorem, T is strongly decomposable. Consequently, T has property (κ) and for any $\{G_1,G_2\} \in \mathrm{cov}\,\sigma(T)$, one has $X = X(T,\overline{G}_1) + X(T,\overline{G}_2)$. We shall use these facts without any further reference.

For an oriented Jordan curve L and a point $a \in L$, we use the upper scripts + and - to distinguish between geometric objects which lie on the positive side from those which lie on the negative side of L, with respect to a.

17.20. THEOREM. Let $T \in B(X)$ have the spectrum on a smooth oriented Jordan curve L. Suppose that, for every $a \in L$, the following conditions are satisfied:

(i) there exists a function $f_a^{\pm}(z)$, locally analytic for $z \neq a$ and $f_a^{\pm}(z) \neq 0$;

(ii) there exist piecewise smooth arcs ℓ_a^+, ℓ_a^- such that $\ell_a^{\pm} \cap L = \{a\}$; $f_a^+(z)$ is bounded in a small area δ^+, $f_a^-(z)$ is bounded in a small area δ^-, with the two areas separated by the arcs ℓ_a^+ and ℓ_a^- as indicated in Fig.1.

(iii) There is $M > 0$, independent of $z \in \ell_a^{\pm} - \{a\}$, such that

$$\|f_a^{\pm}(z)(z-T)^{-1}\| \leq M, \quad \text{for} \quad z \in \ell_a^{\pm} - \{a\}.$$

Then T is decomposable with respect to the identity.

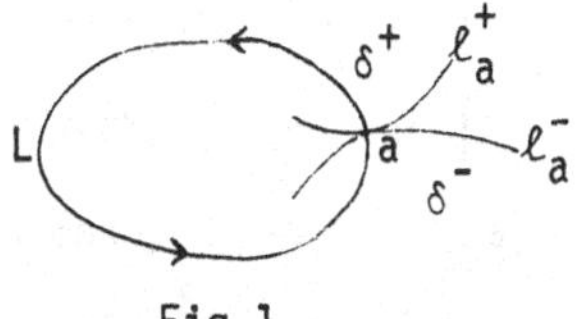

Fig.1.

PROOF. We confine the proof to a closed curve L, that of a nonclosed curve being similar. We shall run the proof in four parts.

Part I. For every pair of disjoint arcs $[a_1,b_1]$ and $[a_2,b_2]$ on L, we draw a piecewise smooth Jordan curve Γ (see Fig.2) with the following properties:

(a) $\Gamma \cap L = \{a_1,b_1,a_2,b_2\}$ and $\ell_{a_1}^- \cup \ell_{b_1}^+ \cup \ell_{a_2}^- \cup \ell_{b_2}^+ \subset \Gamma$ (under the assumption that the arcs $\ell_{a_1}^-, \ldots, \ell_{b_2}^+$ are very small);

(b) $(b_1,a_2) \cup (b_2,a_1)$ is inside Γ.

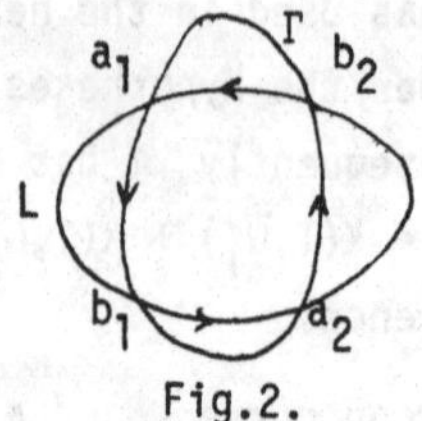

Fig.2.

Put

$$\Phi(\lambda) = f^{-}_{a_1}(\lambda)f^{+}_{b_1}(\lambda)f^{-}_{a_2}(\lambda)f^{+}_{b_2}(\lambda)$$

and define the operator

$$A = \frac{1}{2\pi i}\int_{\Gamma} \Phi(\mu)(\mu-T)^{-1}d\mu.$$

Note that A is bounded and commutes with T. For every $\lambda \notin [b_1,a_2]\cup[b_2,a_1]$, we can choose Γ so that λ is outside Γ. Then, for every $x \in X$,

$$(\lambda-T)\left[\frac{1}{2\pi i}\int_{\Gamma} \frac{\Phi(\mu)(\mu-T)^{-1}x}{\lambda-\mu}\,d\mu\right] = Ax$$

and hence

$$(17.41)\qquad Ax \in X(T,[b_1,a_2] \cup [b_2,a_1]).$$

Recall that, for F closed, X(T,F) is closed.

Part II. In this part, we shall define $Q \in B(X)$ such that, for each $x \in X$,

$$(17.42)\qquad Qx = \begin{cases} x, & \text{if } \sigma(x,T) \subset (a_1,b_1); \\ 0, & \text{if } \sigma(x,T) \subset (a_2,b_2). \end{cases}$$

For every $\lambda \notin [a_1,b_1] \cup [a_2,b_2]$, we can choose Γ so that λ is inside Γ. Put

$$R_\lambda = \frac{1}{2\pi i}\frac{1}{\Phi(\lambda)}\int_{\Gamma} \frac{\Phi(\mu)(\mu-T)^{-1}}{\mu-\lambda}\,d\mu$$

and note that R_λ depends on λ, only. For $x \in X$ with $\sigma(x,T) \subset (a_1,b_1) \cup (a_2,b_2)$, it follows from (17.41) that

$$Ax \in X(T,[b_1,a_2] \cup [b_2,a_1]) \cap X[T,(a_1,b_1) \cup (a_2,b_2)] = \{0\}$$

and hence $Ax = 0$. Consequently, we have

$$(\lambda-T)R_\lambda x = \frac{1}{2\pi i}\frac{1}{\Phi(\lambda)}\int_{\Gamma} \frac{\Phi(\mu)}{\mu-\lambda}\,x\,d\mu = x.$$

For $x \notin X[T,(a_1,b_1) \cup (a_2,b_2)]$, its local resolvent $x(\cdot)$ verifies equality

$$x(\lambda) = R_\lambda x. \tag{17.43}$$

Let γ be a piecewise smooth oriented closed Jordan curve such that $[a_1,b_1]$ is inside γ and $[a_2,b_2]$ is outside γ. Put

$$Q = \frac{1}{2\pi i}\int_\gamma R_\lambda d\lambda.$$

It follows from (17.43) that

$$Qx = \frac{1}{2\pi i}\int_\gamma R_\lambda x d\lambda = \frac{1}{2\pi i}\int_\gamma x(\lambda)d\lambda .$$

Then Q verifies (17.42). Clearly, Q commutes with T. We say that Q is associated to $[a_1,b_1]$.

Part III. Let F_1 and F_2 be closed disjoint sets. In this part, we show that there exists $P \in CM(T)$ such that, for every $x \in X$,

$$Px = \begin{cases} x, & \text{if } \sigma(x,T) \subset F_1; \\ 0, & \text{if } \sigma(x,T) \subset F_2. \end{cases} \tag{17.44}$$

Let G be open such that $F_2 \subset G$, $F_1 \cap \overline{G} = \emptyset$. Clearly, there exists a finite system of arcs $\{(a_j,b_j)\}_{j=1}^n$ such that

$$F_2 \cap L \subset \bigcup_{j=1}^n (a_j,b_j) \subset G \cap L.$$

We may assume that $(a_1,b_1),\dots,(a_n,b_n)$ are arranged in a sequel following the positive orientation of Γ and that they have no endpoint in common. It follows from Part II that, for every $[a_j,b_j]$, there exists $Q_j \in B(X)$, such that

$$Q_j x = \begin{cases} x, & \text{if } \sigma(x,T) \subset (b_j,a_j); \\ 0, & \text{if } \sigma(x,T) \subset (a_j,b_j) \cap F_2. \end{cases} \tag{17.45}$$

Then $P = \prod_{j=1}^n Q_j$ satisfies (17.44) and commutes with T. Indeed, since

$$F_1 \cap L \subset \bigcap_{j=1}^n (b_j,a_j),$$

it follows from (17.45) that

$$Px = \prod_{j=1}^n Q_j x = x, \quad \text{for} \quad x \in X(T,F_1) \tag{17.46}$$

Furthermore, it follows from

$$F_2 \cap L = \bigcup_{j=1}^n [(a_j,b_j) \cap F_2]$$

that $F_2 \cap L$ is the union of n nonintersectiong closed sets

$$F_2 \cap L = \bigcup_{j=1}^{n} K_j ,$$

where $K_j = F_2 \cap (a_j,b_j)$, $1 \le j \le n$. In view of 5.12, we have

$$X(T,F_2) = X(T,F_2 \cap L) = \bigoplus_{j=1}^{n} X(T,K_j),$$

and hence $x \in X(T,F_2)$ has the representation

$$x = \sum_{j=1}^{n} x_j \quad \text{with} \quad x_j \in X(T,K_j), \quad 1 \le j \le n.$$

It follows from (17.45) that $Q_j x_j = 0$, for $1 \le j \le n$, and hence $Px = 0$. Thus, in view of (17.46), P satisfies (17.44).

Part IV. In this final part, we show that T is decomposable with respect to the identity. Let $\{G_1,G_2\} \in \mathbb{C}$ and choose open sets H_1, H_2 such that $\overline{H}_1 \subset G_1$, $\{H_1,G_2\} \in \text{cov}\,\mathbb{C}$, $\overline{H}_2 \subset \overline{G}_2$, $\overline{H}_1 \cap \overline{H}_2 = \emptyset$, $\{H_2,G_1\} \in \text{cov}\,\mathbb{C}$. It follows from Part III that there exists $P \in CM(T)$ satisfying conditions

$$\text{(17.47)} \qquad Px = \begin{cases} x, & \text{if } \sigma(x,T) \subset \overline{H}_1, \\ 0, & \text{if } \sigma(x,T) \subset \overline{H}_2, \end{cases}$$

for all $x \in X$. Put $P_1 = P$, $P_2 = I - P$ and note that

$$\text{(17.48)} \qquad R(P_j) \subset X(T,\overline{G}_j), \quad j=1,2.$$

In fact, $\{G_1,H_2\} \in \text{cov}\,\mathbb{C}$ implies that

$$X = X(T,\overline{G}_1) + X(T,\overline{H}_2)$$

and hence $x \in X$ has a representation

$$x = x_1 + x_2 \quad \text{with} \quad x_1 \in X(T,\overline{G}_1),\ x_2 \in X(T,\overline{H}_2).$$

It follows from (17.47) that

$$P_1 x = P_1 x_1 + P_1 x_2 = P_1 x_1 \in X(T,\overline{G}_1)$$

and, similarly, $P_2 x \in X(T,\overline{G}_2)$. It follows that (17.48) holds; thus P_1, P_2 satisfy the defining conditions of 17.1 and hence those of 17.19. □

NOTES AND COMMENTS

The strong spectral decomposition property, with the implications 15.2-15.8, appeared in [W-Sun]. It represents a generalization of Apostol's concept of strongly decomposable operator [Ap.1968,a]. An extension of the strongly decomposable operator concept to the unbounded case is given in [Ho.1983]. Properties 15.7 and 15.8, confined to bounded strongly decomposable operators, together with the results expressed by 15.10-15.13, appeared in [E-W.1984]. Theorem 15.14 is excerpted from [W,a] and Corollary 15.15 originates from [W-Li.1984].

The strong spectral resolvent concept appeared in [W-E.1983,a], as a special case of the almost localized spectral resolvent [E-W.1983]. The latter is a generalization of the *almost localized spectrum* concept [V.1971,a]. It is worth mentioning that the almost localized spectrum has been used in [Fr.1977], and ultimately lead Radjabalipour to the proof of the equivalence of decomposable and 2-decomposable operators [R.1978]. In this text, the almost localized spectrum appeared in a slightly different form, as the key to the proof of the equivalence of the SDP and the 1-SDP, for unbounded closed operators (Theorem 6.2). Properties 16.2-16.5 were published in [W-E.1983]. Corollary 16.6 is a variant of a property from [B.1973].

Section 17 is based on [W.1983] and [W]. As mentioned in the text, the application 17.20 used some results from [R.1974].

APPENDIX

A. THE (**)-VERSION OF THE PREDUAL THEOREM.

It was mentioned in the Notes and Comments on Chapter III that, in a former version of the predual theorem, the authors used the (**)-density condition. Since the proof of that variant of the theorem is completely different from the proof of 9.6 and it involves some other aspects of the spectral duality problem, we present in the sequel the (**)-version of the predual theorem.

We introduce a convenient notation for the various topologies involved in the duality theory. If A and B are dual spaces, we write $\tau(A,B)$ for the topology on A induced by B, under the given duality. In this vein, $\tau(X,X^*)$ is the weak topology on X and $\tau(X^*,X)$ is the weak*-topology of X^*.

A.1. LEMMA. Let Y_1 be a subspace of X^*. Then Y_1 is closed for $\tau(X^*,X)$ iff there exists a subspace Y_3 of X^{***}, closed for $\tau(X^{***},X^{**})$, such that

(i) Y_3 is invariant under the projection P of X^{***} onto KX^*, along $(JX)^a$;

(ii) interpreting Y_1 as a subspace of X^{***}, under the embedding K,

$$Y_1 = PY_3.$$

PROOF. (If): Let S_3 be the closed (in the metric topology) unit ball of PY^{***} and let $\{x_\alpha^{***}\} \subset S_3$ be a net converging to $x_0^{***} \in KX^*$ for $\tau(KX^*,JX)$. Since $\{x_\alpha^{***}\}$ is bounded in X^{***}, there is a subnet $\{x_\beta^{***}\}$ of $\{x_\alpha^{***}\}$ such that $x_\beta^{***} \to x^{***} \in X^{***}$ for $\tau(X^{***},X^{**})$. Since, by hypothesis, Y_3 is closed for $\tau(X^{***},X^{**})$, we have $x^{***} \in Y_3$. Let $Jx \in JX$. Then

$$\langle Jx,(I-P)x^{***}\rangle = 0$$

and hence

$$\lim_\beta \langle Jx,x_\beta^{***}\rangle = \langle Jx,x^{***}\rangle = \langle Jx,Px^{***}\rangle . \tag{A.1}$$

On the other hand, we have

$$\lim_\beta \langle Jx,x_\beta^{***}\rangle = \lim_\alpha \langle Jx,x_\alpha^{***}\rangle = \langle Jx,x_0^{***}\rangle. \tag{A.2}$$

Thus $\langle Jx,Px^{***}\rangle = \langle Jx,x_0^{***}\rangle$, by (A.1) and (A.2). Since both Px^{***} and x_0^{***} are elements of KX^*, it follows that

$$x_0^{***} = Px^{***} \in PY_3. \tag{A.3}$$

Since clearly, $\|x_0^{***}\| \leq 1$, (A.3) implies that $x_0^{***} \in S_3$ and hence S_3 is closed for $\tau(KX^*,JX)$. By the Kreĭn-Šmulian theorem, PY_3 is closed for $\tau(KX^*,JX)$. In view of (ii), Y_1 is closed for $\tau(X^*,X)$.

(Only if): $Y_3 = Y_1^{aa}$ is closed for $\tau(X^{***},X^{**})$. For simplicity, we consider Y_1 and KY_1 at the same level, i.e. as subspaces of X^{***}, and we shall extend similar considerations to other subspaces at appropriate times. Y_1^a being the annihilator of Y_1 in X^{**} and since Y_1 is a subspace of X^{***}, Y_3 is the $\tau(X^{***},X^{**})$-closure of Y_1 (see e.g. [Br-Pea.1977, Proposition 16.2]). Then, for every $x^{***} \in Y_3$, there exists a net $\{x_\alpha^*\} \subset Y_1$ such that $x_\alpha^* \to x^{***}$ for $\tau(X^{***},X^{**})$. Hence, for every $x \in X$, it follows from $X^a = (I-P)X^{***}$ that

$$\langle x,Px^{***}\rangle = \langle x,Px^{***}\rangle + \langle x,(I-P)x^{***}\rangle$$

$$= \langle x,x^{***}\rangle = \lim_\alpha \langle x,x_\alpha^*\rangle ,$$

where, as mentioned earlier, we regard X as a subspace of X^{**}. Thus, $\{x_\alpha^*\}$, as a net in X^*, converges to Px^{***} for $\tau(X^*,X)$. Since, by hypothesis, Y_1 is closed for $\tau(X^*,X)$, $\{x_\alpha^*\} \subset Y_1$ implies that $Px^{***} \in Y_1$ and hence $PY_3 \subset Y_1$. It follows from the evident equality $PY_1 = Y_1$ that $PY_3 = Y_1$ and $Y_3 \in \operatorname{Inv} P$. □

A.2. LEMMA. Suppose that the subspace $Y_1 \subset X^*$ is closed for $\tau(X^*,X)$. Then, with Y_3 as defined in A.1, we have

$$(X^*/Y_1)^{**} \cong (X^*/Y_1) \oplus (X^a/X^a \cap Y_3), \tag{A.4}$$

where $\cong$ denotes an equivalence, under topological isomorphism.

PROOF. In view of A.1, we have

$$Y_3 = Y_1 \oplus (I-P)Y_3,$$

and then $(I-P)Y_3 = X^a \cap Y_3$ implies

$$Y_3 = Y_1 \oplus (X^a \cap Y_3). \tag{A.5}$$

Put $Y_2 = Y_1^a$. The equivalences

$$(X^*/Y_1)^* \cong Y_2, \quad X^{***}/Y_3 \cong Y_2^*,$$

under isometric isomorphism, being known (e.g. [Br-Pea.1977; Propositions 16.6 and 16.7]), we have

$$(X^*/Y_1)^{**} \cong X^{***}/Y_3.$$

Thus, to obtain (A.4), we have to show that

$$\text{(A.6)} \qquad X^{***}/Y_3 \cong (X^*/Y_1) \oplus (X^a/X^a \cap Y_3).$$

Recall that the norm of the direct sum of two Banach spaces A and B is defined by $\|\cdot\|_{A\oplus B} = \|\cdot\|_A + \|\cdot\|_B$. Define the mapping

$$\Phi : X^{***}/Y_3 \to (X^*/Y_1) \oplus (X^a/X^a \cap Y_3)$$

by

$$\text{(A.7)} \qquad \Phi(x^{***} + Y_3) = (x_1^{***} + Y_1) \oplus [x_2^{***} + (X^a \cap Y_3)]$$

with $x^{***} \in X^{***}$, $x_1^{***} \in X^*$, $x_2^{***} \in X^a$ and $x^{***} = x_1^{***} + x_2^{***}$. It is easily seen that Φ is a well-defined surjection. To see that Φ is injective, suppose that there exists $z^{***} \in X^{***}$ such that

$$\Phi(z^{***} + Y_3) = \Phi(x^{***} + Y_3).$$

Then, for $z_1^{***} \in X^*$, $z_2^{***} \in X^a$ with $z^{***} = z_1^{***} + z_2^{***}$, we have

$$(z_1^{***} + Y_1) \oplus [z_2^{***} + (X^a \cap Y_3)] = (x_1^{***} + Y_1) \oplus [x_2^{***} + (X^a \cap Y_3)]$$

and hence (A.5) implies that $z^{***} + Y_3 = x^{***} + Y_3$. Thus Φ is injective. Furthermore, it follows from the inequality

$$\begin{aligned} \|x^{***} + Y_3\| &= \inf\{\|x^{***} + y^{***}\| : y^{***} \in Y_3\} \\ &\le \inf\{\|x_1^{***} + y_1^{***}\| : y_1^{***} \in Y_1\} \\ &\quad + \inf\{\|x_2^{***} + y_2^{***}\| : y_2^{***} \in X^a \cap Y_3\} \end{aligned}$$

and from the open mapping theorem that Φ is a topological mapping. Thus (A.6) holds and hence (A.4) follows. □

For $Y_1 \subset X^*$ closed for $\tau(X^*,X)$, the annihilator $Y_2 = Y_1^a$ of Y_1 in X^{**} and the preannihilator $Y_0 = {}^aY_1$ of Y_1 in X give rise to some further equivalences, under isometric isomorphism, and solicit some additional notational conventions. In fact, by defining

$$Y_0 = {}^aY_1, \quad Y_2 = Y_1^a \quad \text{and} \quad Y_1^{aa} = Y_2^a,$$

we have $Y_2 \cong Y_0^{**}$ (e.g. [Br-Pea.1977, Propositions 16.6, 16.7]) and

$$Y_2^* \cong X^{***}/Y_3.$$

Moreover, the first, the second and the third dual of the quotient space X/Y_0 is equivalent, under isometric isomorphism, to Y_1, X^{**}/Y_2 and Y_3, respectively. Therefore, Y_0 can be regarded as a subspace of Y_2 and X/Y_0 as a subspace of X^{**}/Y_2. We shall distinguish between the annihilator Y_0^a ($\subset X^*$) of Y_0 as a subspace of X and the annihilator $Y_0^\alpha(\subset Y_2^*)$ of Y_0 as a subspace of Y_2. $(X/Y_0)^\alpha(\subset Y_3)$ annihilates X/Y_0 as a subspace of X^{**}/Y_2.

A.3. LEMMA. Suppose that the subspace $Y_1 \subset X^*$ is closed for $\tau(X^*,X)$ and $Y_3 = Y_1^{aa}$. Then

(i) $Y_0^{\alpha} \cong X^a/X^a \cap Y_3$;

(ii) $(X/Y_0)^{\alpha} \cong X^a \cap Y_3$;

(iii) the coinduced P/Y_3 on X^{***}/Y_3 acts as a projection onto X^*/Y_1 along $X^a/X^a \cap Y_3$.

PROOF. For $x^{***} \in X^{***}$, let $(x^{***})\hat{} = x^{***} + Y_3$;
for $\xi^{***} \in X^*$ $(\subset X^{***})$, let $(\xi^{***})^{\sim} = \xi^{***} + Y_1$;
for $\zeta^{***} \in X^a$ $(\subset X^{***})$, let $(\zeta^{***})^{\backsim} = \zeta^{***} + (X^a \cap Y_3)$.

(i): As mentioned earlier, X^*/Y_1 and X^{***}/Y_3 are regarded as the first and the third dual of Y_0, respectively. Let $x \in Y_0$. Since $Y_0 \subset Y_2$ and $Y_2^a = Y_3$, it follows from (A.7) that, for $x_2^{***} \in X^a$,

$$\Phi[(x_2^{***})\hat{}] = 0 \oplus (x_2^{***})^{\backsim}.$$

Then

$$\langle x,(x_2^{***})\hat{}\rangle = \langle x,(x_2^{***})^{\backsim}\rangle = \langle x,x_2^{***} + y^{***}\rangle = 0,$$

where $y^{***} \in X^a \cap Y_3$. Thus, we have $(x_2^{***})\hat{} \in Y_0^{\alpha}$.

Conversely, let $(x^{***})\hat{} \in Y_0^{\alpha}$. In view of (A.7),

$$\Phi[(x^{***})\hat{}] = (x_1^{***})^{\sim} \oplus (x_2^{***})^{\backsim},$$

where $(x_1^{***})^{\sim} = x_1^{***} + Y_1 \in X^*/Y_1$, $(x_2^{***})^{\backsim} = x_2^{***} + (X^a \cap Y_3)$, and $x_2^{***} = x^{***} - x_1^{***}$. Then, for $x \in Y_0$,

$$\langle x,(x_2^{***})^{\backsim}\rangle = 0 \text{ and } \langle x,(x^{***})\hat{}\rangle = 0$$

imply that

$$\langle x,(x_1^{***})^{\sim}\rangle = 0.$$

Since $Y_0^* \cong X^*/Y_1$, $(x_1^{***})^{\sim} \in X^*/Y_1$, we have $(x_1^{***})^{\sim} = 0$ and hence

$$\Phi[(x^{***})\hat{}] = 0 \oplus (x_2^{***})^{\backsim} \in X^a/X^a \cap Y_3.$$

Thus, by having seen that $(x^{***})\hat{} \in Y_0^{\alpha}$ iff $\Phi[(x^{***})\hat{}] = 0 \oplus (x_2^{***})^{\backsim}$, the proof of (i) is concluded.

(ii): Let $x \in X$, $y \in Y_0$ and $\mathring{x} = x + Y_0 \in X/Y_0$. For $y^{***} \in X^a \cap Y_3$, we have

$$\langle \mathring{x},y^{***}\rangle = \langle x + y,y^{***}\rangle = 0$$

and since $(X/Y_0)^{***} \cong Y_3$, it follows that $y^{***} \in (X/Y_0)^{\alpha}$.

Conversely, let $y^{***} \in Y_3$ be such that $<\overset{\circ}{x},y^{***}> = 0$. In view of (A.5), there is a representation

$$y^{***} = y_1^{***} + y_2^{***}, \quad y_1^{***} \in Y_1, \quad y_2^{***} \in X^a \cap Y_3.$$

It follows from $<\overset{\circ}{x},y_2^{***}> = 0$, that $<\overset{\circ}{x},y_1^{***}> = 0$. Thus

$$<x,y_1^{***}> = <x+y,y_1^{***}> = <\overset{\circ}{x},y_1^{***}>$$

implies that $<x,y_1^{***}> = 0$ and hence $y_1^{***} = 0$. Consequently, $y^{***} = y_2^{***} \in X^a \cap Y_3$. The proof of (ii) is thus concluded, by having seen that $y^{***} \in (X/Y_0)^\alpha$ iff $y^{***} \in X^a \cap Y_3$.

(iii): Since $Y_3 \in \text{Inv}\, P$, it follows from A.1 that the co-induced operator P/Y_3 exists on X^{***}/Y_3. Evidently, P/Y_3 is a projection. Let $(x_1^{***})^\wedge$ be such that

$$\Phi[(x_1^{***})^\wedge] \in X^*/Y_1.$$

Without loss of generality, we may assume that the element x_1^{***} is in X^*. Then we have

$$(P/Y_3)(x_1^{***})^\wedge = (Px_1^{***})^\wedge = x_1^{***} .$$

Similarly, for $(x_2^{***})^\wedge$ with

$$\Phi[(x_2^{***})^\wedge] = 0 \oplus (x_2^{***})^\sim,$$

we obtain

$$(P/Y_3)(x_2^{***})^\wedge = 0.$$

It follows from the direct sum (A.6) that P/Y_3 acts as a projection onto X^*/Y_1 along $X^a/X^a \cap Y_3$. □

A.4. LEMMA. Given T, assume that (**) holds and the subspaces Y_i $(0 \le i \le 3)$ are, as defined in A.1 - A.3.

(i) Assume that

$Y_1 \in \text{Inv}\, T^*$ is closed for $\tau(X^*,X)$; $T^*|Y_1$ is densely defined and T^*/Y_1 is bounded on X^*/Y_1.

Then

$Y_2 \subset \mathcal{D}_{T^{**}}$, $Y_2 \in \text{Inv}\, T^{**}$ and $Y_3 \in \text{Inv}\, T^{***}$.

(ii) If, in addition, we assume that

$\mathcal{D}(T^{***}/Y_3) = X^{***}/Y_3$,

then $Y_0 \subset \mathcal{D}_T$ and $Y_0 \in \text{Inv}\, T$.

PROOF. (i). Given $x^* \in X^*$, let $(x^*)^\sim = x^* + Y_1 \in X^*/Y_1$. For every $x^* \in \mathcal{D}_{T^*}$, $y^* \in Y_1$ and $y^{**} \in Y_2$, we have

$$|<T^*x^*,y^{**}>| = |<T^*x^* + y^*,y^{**}>| = |<(T^*/Y_1)(x^*)^\sim,y^{**}>|$$
$$\leq \|(T^*/Y_1)\| \cdot \|(x^*)^\sim\| \cdot \|y^{**}\| \leq \|(T^*/Y_1)\| \cdot \|x^*\| \cdot \|y^{**}\|.$$

Hence, for fixed $y^{**} \in Y_2$, $<T^*x^*,y^{**}>$ is a bounded linear functional of x^*. Therefore, $y^{**} \in \mathcal{D}_{T^{**}}$ or, equivalently, $Y_2 \subset \mathcal{D}_{T^{**}}$. Furthermore, for every $y^* \in Y_1 \cap \mathcal{D}_{T^*}$, it follows from

$$<y^*,T^{**}y^{**}> = <T^*y^*,y^{**}> = 0$$

and from $\overline{Y_1 \cap \mathcal{D}_{T^*}} = Y_1$, that $T^{**}y^{**} \in Y_2$ and hence $Y_2 \in \text{Inv } T^{**}$. In a similar way, one can show that $Y_3 \in \text{Inv } T^{***}$.

(ii). Since $T^{**}|Y_2$ is bounded by (i), its conjugate $(T^{**}|Y_2)^*$ is bounded on X^{***}/Y_3. Next, we are going to show that

$$\text{(A.8)} \qquad T^{***}/Y_3 = (T^{**}|Y_2)^*.$$

Given $y^{***} \in X^{***}$, we denote by $(y^{***})^\wedge = y^{***} + Y_3 \in X^{***}/Y_3$. It follows from the hypothesis that, for every $(y^{***})^\wedge \in X^{***}/Y_3$, there exists $z^{***} \in (y^{***})^\wedge \cap \mathcal{D}_{T^{***}}$. Then, for every $y^{**} \in Y_2$, we have successively

$$<y^{**},(T^{**}|Y_2)^*(y^{***})^\wedge> = <T^{**}y^{**},(y^{***})^\wedge> = <T^{**}y^{**},(z^{***})^\wedge>$$
$$= <T^{**}y^{**},z^{***}> = <y^{**},T^{***}z^{***}> = <y^{**},(T^{***}/Y_3)(z^{***})^\wedge>$$
$$= <y^{**},(T^{***}/Y_3)(y^{***})^\wedge>.$$

The equality (A.8) is thus obtained.

By 10.4, the projection P (of X^{***} onto KX^*, along $(JX)^a$) commutes with T^{***}. Then P/Y_3 commutes with $(T^{**}|Y_2)^*$. It follows from A.2, A.3(iii) and A.8 that $X^a/X^a \cap Y_3$ is invariant under $(T^{**}|Y_2)^*$ and $X^a/X^a \cap Y_3 \subset \mathcal{D}[(T^{**}|Y_2)^*]$. Let $y \in Y_0$ and $(y^{***})^\sim \in X^a/X^a \cap Y_3$. Since $Y_0 \subset Y_2 \subset \mathcal{D}_{T^{**}}$, it follows from A.2 and A.3 (i), that

$$<T^{**}y,(y^{***})^\sim> = <y,(T^{**}|Y_2)^*(y^{***})^\sim> = 0.$$

Thus $T^{**}y \in Y_0$ and hence $Y_0 \in \text{Inv } T^{**}$.

For every $y \in Y_0$ and $y^* \in \mathcal{D}_{T^*}$, it follows from

$$<T^*y^*,y> = <y^*,T^{**}y>,$$

and from $T^{**}y \in Y_0 \subset X$, that

$$(y, T^{**}y) \in {}^{a}[VG(T^*)] = G(T).$$

Thus, we have $y \in \mathcal{D}_T$ and $Ty = T^{**}y$ which implies $Y_0 \subset \mathcal{D}_T$ and $Y_0 \in \mathrm{Inv}\, T$. □

A.5. LEMMA. Given T, assume that condition (**) holds and let Y_i $(0 \leq i \leq 3)$ be as in the foregoing lemmas. If $Y_1 \in \mathrm{Inv}\, T^*$ is closed for $\tau(X^*, X)$ and there is $Z_1 \in \mathrm{Inv}\, T^*$ such that

$$X^* = Y_1 + Z_1, \quad Z_1 \subset \mathcal{D}_{T^*}, \tag{A.9}$$

then Y_1 and Y_3 satisfy the hypotheses of the assertions (i) and (ii) of A.4. In particular, $Y_0 \subset \mathcal{D}_T$ and $Y_0 \in \mathrm{Inv}\, T$.

PROOF. In view of 3.4 and 5.15, Y_1 satisfies the conditions stated in A.4 (i). Put $Z_3 = Z_1^{aa}$. Since Z_1 may be regarded as a subspace contained in X^{***}, Z_3 is the closure of Z_1 for $\tau(X^{***}, X^{**})$, (e.g. [Br-Pea.1977, Proposition 16.2]). We shall run the remainder of the proof in three parts.

Part I. In this part, we shall obtain the decomposition

$$X^{***} = Y_3 + Z_3. \tag{A.10}$$

Since the unit ball of X^* is dense in the unit ball of X^{***} with respect to $\tau(X^{***}, X^{**})$, for a given $x^{***} \in X^{***}$, there is a net $\{x_\alpha^*\} \subset X^*$ such that

$$\|x_\alpha^*\| \leq \|x^{***}\| \quad \text{and} \quad x_\alpha^* \to x^{***} \quad \text{for} \quad \tau(X^{***}, X^{**}).$$

In view of (A.9), for every α, there is a representation

$$x_\alpha^* = x_{\alpha 1}^* + x_{\alpha 2}^*, \quad x_{\alpha 1}^* \in Y_1, \quad x_{\alpha 2}^* \in Z_1 \tag{A.11}$$

and there is a number $M > 0$ such that

$$\|x_{\alpha 1}^*\| + \|x_{\alpha 2}^*\| \leq M \|x_\alpha^*\|.$$

Then nets $\{x_{\alpha i}^*\}$ $(i=1,2)$ are bounded and hence, for $i=1,2$, there exists a subnet $\{x_{\beta i}^*\}$ of $\{x_{\alpha i}^*\}$ converging to a point x_i^{***} for $\tau(X^{***}, X^{**})$. Y_3 and Z_3 being closed for $\tau(X^{***}, X^{**})$, we have $x_1^{***} \in Y_3$, $x_2^{***} \in Z_3$. By substituting β for α in (A.11), one obtains

$$x^{***} = \lim_\beta x_\beta^* = \lim_\beta x_{\beta 1}^* + \lim_\beta x_{\beta 2}^* = x_1^{***} + x_2^{***}$$

and this gives rise to the decomposition (A.10).

Part II. In this part, we show that

$$Z_3 \subset \mathcal{D}_{T^{***}} \quad \text{and} \quad Z_3 \in \mathrm{Inv}\, T^{***}. \tag{A.12}$$

Putting $Z_2 = Z_1^a$, it follows from the hypotheses $Z_1 \subset \mathcal{D}_{T^*}$ and $Z_1 \in \text{Inv}\, T^*$, that $Z_2 \in \text{Inv}\, T^{**}$. Furthermore, since T^{**} is densely defined, so is T^{**}/Z_2. On the other hand, $T^*|Z_1$ is bounded and hence $(T^*|Z_1)^*$ is bounded on X^{**}/Z_2. One can easily see that $(T^*|Z_1)^*$ is a closed (actually bounded) extension of T^{**}/Z_2.

Let $x^{***} \in Z_3$. For every $y^{**} \in \mathcal{D}_{T^{**}}$ and $(y^{**})^\wedge = y^{**} + Z_2 \in X^{**}/Z_2$, the equalities

(A.13) $\qquad \langle T^{**}y^{**},x^{***}\rangle = \langle (T^{**}/Z_2)(y^{**})^\wedge,x^{***}\rangle = \langle (T^*|Z_1)^*(y^{**})^\wedge,x^{***}\rangle$

imply

$$|\langle T^{**}y^{**},x^{***}\rangle| \le \|(T^*|Z_1)^*\|\cdot\|(y^{**})^\wedge\|\cdot\|x^{***}\|$$
$$\le \|(T^*|Z_1)^*\|\cdot\|y^{**}\|\cdot\|x^{***}\|.$$

Thus, it follows that $\langle T^{**}y^{**},x^{***}\rangle$ is a bounded linear functional of y^{**}. Consequently, $x^{***} \in \mathcal{D}_{T^{***}}$ and hence $Z_3 \subset \mathcal{D}_{T^{***}}$. It follows from (A.13) that, for $y^{**} \in \mathcal{D}_{T^{**}}$ and $x^{***} \in Z_3$, we have

$$\langle y^{**},T^{***}x^{***}\rangle = \langle T^{**}y^{**},x^{***}\rangle = \langle (T^*|Z_1)^*(y^{**})^\wedge,x^{***}\rangle .$$

Since, by hypothesis, $\overline{\mathcal{D}}_{T^{**}} = X^{**}$ and $(T^*|Z_1)^*$ is bounded, for every $y^{**} \in X^{**}$ and $x^{***} \in Z_3$, we have

$$\langle y^{**},T^{***}x^{***}\rangle = \langle (T^*|Z_1)^*(y^{**})^\wedge,x^{***}\rangle .$$

In particular, for $y^{**} \in Z_2$, $(y^{**})^\wedge = \hat{0}$ implies that

$$\langle y^{**},T^{***}x^{***}\rangle = 0$$

and hence $T^{***}x^{***} \in Z_3$. The assertions (A.12) are thus proved.

Part III. It follows from (A.10), (A.12) and 3.4 that T^{***}/Y_3 is bounded on X^{***}/Y_3. Thus, the hypotheses of assertion (ii) in A.4 are satisfied by Y_3. □

The (**)-version of 9.3 and 9.4 now follows.

A.6. THEOREM. Given T, assume that (**) holds. If T^* has the SDP, then

(I) for every closed F, $X^*(T^*,F)$ is closed for $\tau(X^*,X)$;

(II) for every compact F, $\Xi^*(T^*,F)$ is closed for $\tau(X^*,X)$.

PROOF. We confine the proof to (II). Assuming that T^* has the SDP, the duality theorem 8.1 implies that T^{**} and T^{***} have the SDP. Consequently, $\Xi^{***}(T^{***},F)$ is a T^{***}-bounded spectral maximal space and hence it is closed for $\tau(X^{***},X^{**})$, ([V.1971,a; Proposition 2.9] or [V.1982, IV.

Proposition 5.6]). Next, we prove the equality

(A.14) $$K\Xi^*(T^*,F) = P\Xi^{***}(T^{***},F).$$

Let $x^{***} \in K\Xi^*(T^*,F)$. Since T^* and $T^{***}|KX^*$ are similar (Theorem 10.2 (II)), $T^{***}|KX^*$ has the SDP and $K\Xi^*(T^*,F)$ is a $(T^{***}|KX^*)$-bounded spectral maximal space. Consequently,

(A.15) $$\sigma(x^{***},T^{***}) \subset F \text{ and } \lim_{\lambda\to\infty} x^{***}(\lambda) = 0.$$

Quoting 5.11, $x^{***} \in \Xi^{***}(T^{***},F)$ and hence

(A.16) $$K\Xi^*(T^*,F) = PK\Xi^*(T^*,F) \subset P\Xi^{***}(T^{***},F).$$

Conversely, let $x^{***} \in \Xi^{***}(T^{***},F)$. By 5.11, we recapture conditions (A.15) and then

$$\sigma(Px^{***},T^{***}) \subset F \text{ and } \lim_{\lambda\to\infty} Px^{***}(\lambda) = 0.$$

Since $x^{***}(\lambda) \in \mathcal{D}_{T^{***}}$, one has $Px^{***}(\lambda) \in \mathcal{D}_{T^{***}}$, by 10.4. By 10.3, $Px^{***}(\lambda) \in K\mathcal{D}_{T^*}$ and, for $\lambda \in \rho(x^{***},T^{***})$, we obtain

$$(\lambda-T^*)K^{-1}Px^{***}(\lambda) = K^{-1}(\lambda-T^{***})Px^{***}(\lambda)$$
$$= K^{-1}P(\lambda-T^{***})x^{***}(\lambda) = K^{-1}Px^{***}.$$

Thus, $\sigma(K^{-1}Px^{***},T^*) \subset F$ and since $\lim_{\lambda\to\infty} K^{-1}Px^{***}(\lambda) = 0$, 5.11 implies $K^{-1}Px^{***} \in \Xi^*(T^*,F)$, i.e. $Px^{***} \in K\Xi^*(T^*,F)$. Thus,

$$P\Xi^{***}(T^{***},F) \subset K\Xi^*(T^*,F)$$

and this coupled with (A.16) produces (A.14). Now A.1 implies that $K\Xi^*(T^*,F)$ is closed for $\tau(KX^*,JX)$ and hence $\Xi^*(T^*,F)$ is closed for $\tau(X^*,X)$. □

A.7. THEOREM. Given T, assume that (**) holds and T^* has the SDP. Let $G \in V_\infty$ be open, $F = G^c$ and $Y_0 = {}^aX^*(T^*,\overline{G})$. Then

(i) $Y_0 \subset \mathcal{D}_T$, $Y_0 \in \text{Inv } T$, $\sigma(T|Y_0) \subset F$;

(ii) T/Y_0 is closable and, for the minimal closed extension $\overline{T/Y_0}$ of T/Y_0, $\sigma(\overline{T/Y_0}) \subset \overline{G}$.

PROOF. Since T^* has the SDP, $X^*(T^*,\overline{G})$ is closed for $\tau(X^*,X)$ and hence $Y_0^a = X^*(T^*,\overline{G})$. Let $H \supset F$ be open and relatively compact. Then $\{G,H\} \in \text{cov } \mathbb{C}$ with $G \in V_\infty$ and the SDP of T^* gives rise to

(A.17) $$X^* = X^*(T,\overline{G}) + \Xi^*(T^*,\overline{H}).$$

For $Y_1 = X^*(T^*,\overline{G})$ and $Z_1 = \Xi^*(T^*,\overline{H})$, the hypotheses of A.5 are satisfied and hence we have $Y_0 \subset \mathcal{D}_T$, $Y_0 \in \text{Inv } T$. Furthermore, it follows from $Z_1 \subset \mathcal{D}_{T^*}$ and 5.15 that $T^*|Y_1 = T^*|X^*(T^*,\overline{G})$ is densely defined. Then, 3.5 implies that T/Y_0 is closable and for its minimal closed extension $\overline{T/Y_0}$, we have

$$(T/Y_0)^* \cong T^*|Y_0^a = T^*|X^*(T^*,\overline{G});$$

$$\sigma(\overline{T/Y_0}) = \sigma[(T/Y_0)^*] = \sigma[T^*|X^*(T^*,\overline{G})] \subset \overline{G}.$$

It remains to show that $\sigma(T|Y_0) \subset F$. Apply 5.16 to T^* and deduce that $T^*/X^*(T^*,\overline{G})$ is bounded and

$$\sigma[T^*/X^*(T^*,\overline{G})] \subset G^c = F. \tag{A.18}$$

Thus, it follows from the equivalence

$$(T|Y_0)^* \cong T^*/X^*(T^*,\overline{G})$$

and from (A.18) that

$$\sigma(T|Y_0) = \sigma[(T|Y_0)^*] \subset F. \quad \square$$

A.8. PROPOSITION. Let U be a densely defined closed operator on X^* with nonvoid resolvent set. There exists a densely defined closed operator V on X with $V^* = U$ iff U^{**} commutes with P.

PROOF. The "only if" part is a consequence of 10.4.

(If): Let $\lambda_0 \in \rho(U)$ and put $A = R(\lambda_0;U)$. Then

$$A \in B(X^*), \quad A^* = R(\lambda_0;U^*) \text{ and } A^{**} = R(\lambda_0,U^{**}).$$

Since, by hypothesis, P commutes with U^{**} and hence P commutes with A^{**}, X^a is invariant under A^{**}. For every $x \in X$ ($\subset X^{**}$) and $x^{***} \in X^a$, $A^{**}x^{***} \in X^a$ implies

$$\langle A^*x,x^{***}\rangle = \langle x,A^{**}x^{***}\rangle = 0$$

and hence $A^*x \in X$ or, equivalently, $X \in \text{Inv } A^*$.

Put $A_0 = A^*|X \in B(X)$. It follows from $[VG(A)]^a = G(A^*)$, that

$${}^a[VG(A)] = (X \times X) \cap G(A^*) = G(A_0)$$

and hence $A_0^* = A$. Since $\overline{R(A)} = X^*$, A_0 is injective. It follows from $R(A_0)^a = N(A)$, that $\overline{R(A_0)} = X$.

The operator $V = \lambda_0 - A_0^{-1}$ is densely defined on X and the equalities

$$[(\lambda_0-V^*)]^{-1} = [(\lambda_0-V)^{-1}]^* = A_0^* = A = R(\lambda_0,U)$$

imply that $(\lambda_0'-V)^* = \lambda_0-U$ and hence $V^* = U$. □

A.9. THEOREM. Given T, assume that (**) holds and T* has the SDP. Let $G \in G^K$, $F = G^c$ and $Y_0 = {}^a\Xi^*(T^*,\overline{G})$. Then

$$Y_0 \in \text{Inv } T \quad \text{and} \quad \sigma(T|Y_0) \subset F.$$

PROOF. Put $Y_1 = \Xi^*(T^*,\overline{G})$, $Y_2 = Y_1^a$, $Y_3 = Y_1^{aa}$, denote by Y_0^α the annihilator of $Y_0(\subset Y_2)$ in X^{***}/Y_3 and by $(X/Y_0)^\alpha$ the annihilator of $X/Y_0(\subset X^{**}/Y_2)$ in Y_3. By A.3, P/Y_3 acts on X^{***}/Y_3 as a projection onto X^*/Y_1 along $X^a/X^a \cap Y_3$. Since P commutes with T*** (Lemma 10.4), P/Y_3 commutes with the coinduced T^{***}/Y_3 on X^{***}/Y_3. Since T* is densely defined, so is T^*/Y_1 on X^*/Y_1.

Apply 5.15 to T**, which has the SDP by 8.1, and infer that $T^{**}|Y_2$ is densely defined. On the other hand, 5.16 applied to T* asserts that T^*/Y_1 is closed and $\sigma(T^*/Y_1) \subset G^c$. Hence $\rho(T^*/Y_1) \neq \emptyset$. It is easily seen that the first and the second conjugate of T^*/Y_1 is $T^{**}|Y_2$ and T^{***}/Y_3, respectively. Then $U = T^*/Y_1$ satisfies the hypotheses of A.8 asserting that there is a closed operator V, densely defined on Y_0 such that $V^* = T^*/Y_1$. We are going to show that $V = T|Y_0$.

First, $Y_0 \in \text{Inv } T$. In fact, for $x \in Y_0 \cap \mathcal{D}_T$, $x^* \in Y_1$, we have

$$\langle Tx,x^*\rangle = \langle x,T^*x^*\rangle = 0$$

and hence $Tx \in Y_0$ or, equivalently, $Y_0 \in \text{Inv } T$.

Next,

$$G(V) = (Y_0 \times Y_0) \cap G(T). \tag{A.19}$$

Indeed, for $(x^*)^\wedge \in \mathcal{D}_{T^*/Y_1}$, $x^* \in (x^*)^\wedge$ and $x \in Y_0 \cap \mathcal{D}_T$, we have

$$\langle x,(T^*/Y_1)(x^*)^\wedge\rangle = \langle x,T^*x^*\rangle = \langle Tx,x^*\rangle = \langle Tx,(x^*)^\wedge\rangle$$

and hence

$$(Y_0 \times Y_0) \cap G(T) \subset {}^a[VG(T^*/Y_1)] = G(V). \tag{A.20}$$

On the other hand, for $(x,y) \in Y_0 \times Y_0$ satisfying

$$(x,y) \in {}^a[VG(T^*/Y_1)] = G(V),$$

and, for every $x^* \in \mathcal{D}_{T^*}$, we have

$$\langle x,T^*x^*\rangle = \langle x,(T^*/Y_1)(x^*)^\wedge\rangle = \langle Vx,(x^*)^\wedge\rangle = \langle Vx,x^*\rangle = \langle y,x^*\rangle.$$

Thus, it follows that $(x,y) \in {}^a[VG(T^*)] = G(T)$ or, equivalently,

(A.21) $$(Y_0 \times Y_0) \cap G(T) \supset G(V).$$

Now (A.20) and (A.21) imply (A.19) and it follows from the latter that

$$(T|Y_0)^* = T^*/Y_1.$$

Finally, apply 5.16 to T^*/Y_1 and conclude the proof, by inferring that

$$\sigma(T|Y_0) = \sigma(T^*/Y_1) \subset G^c = F. \quad \square$$

The (**)-version of the predual theorem now follows.

A.10. THEOREM. Given T assume that (**) holds. If T* has the SDP then T has the SDP.

PROOF. Let $\{G_0,G_1\} \in \operatorname{cov} \mathbb{C}$ with $G_0 \in V_\infty$ and $G_1 \in G^K$. Let F_0, F_1 be closed such that $F_0 \subset G_0$, $F_1 \subset G_1$ and $\{\operatorname{Int} F_0, \operatorname{Int} F_1\} \in \operatorname{cov} \mathbb{C}$. Then, for $H_0 = F_1^c$, $H_1 = F_0^c$, we have $\overline{H}_0 \cap \overline{H}_1 = \emptyset$. Put $Y = {}^aX^*(T^*,\overline{H}_0 \cup \overline{H}_1)$. By A.7, $Y \subset \mathcal{D}_T$, T/Y is closable and, for the minimal closed extension $\overline{T/Y}$ of T/Y, we have

$$\sigma(\overline{T/Y}) \subset \overline{H}_0 \cup \overline{H}_1.$$

Since $\overline{H}_0$, $\overline{H}_1$ are disjoint (with $\overline{H}_1$ compact), it follows from the functional calculus, that there are subspaces $\hat{Z}_0$, $\hat{Z}_1$ of X/Y, invariant under $\overline{T/Y}$, such that

(A.22) $$X/Y = \hat{Z}_0 \oplus \hat{Z}_1.$$

Putting $Z_i = \{x \in X : x \in \hat{x},\ \hat{x} \in \hat{Z}_i\}$, $i=0,1$; (A.22) implies

(A.23) $$X = Z_0 + Z_1.$$

Define $Y_0 = {}^a\Xi^*(T^*,\overline{H}_1)$, $Y_1 = {}^aX^*(T^*,\overline{H}_0)$ and use A.7, A.9, to obtain

(A.24) $$Y_1 \subset \mathcal{D}_T,\ Y_i \in \operatorname{Inv} T,\ \sigma(T|Y_i) \subset (\overline{H}_j)^c \subset F_i \subset G_i,\ i \neq j;\ i,j=0,1.$$

To conclude the proof, we shall show that $Z_i \subset Y_i$, $i=0,1$. Let $x \in Z_0$ and $x^* \in \Xi^*(T^*,\overline{H}_1)$. Then $\hat{x} \in \hat{Z}_0$. Define

(A.25) $$g(\lambda) = \begin{cases} \langle \hat{x}, R[\lambda;T^*|\Xi^*(T^*,\overline{H}_1)]x^*\rangle, & \text{if } \lambda \in (\overline{H}_1)^c; \\ \langle R(\lambda;\overline{T/Y}|\hat{Z}_0)\hat{x},x^*\rangle, & \text{if } \lambda \in (\overline{H}_0)^c. \end{cases}$$

Since $(\overline{T/Y})^* \cong T^*|Y^a = T^*|X^*(T^*,\overline{H}_0 \cup \overline{H}_1)$ and $\Xi^*(T^*,\overline{H}_1) \subset X^*(T^*,\overline{H}_0 \cup \overline{H}_1)$, for $\lambda \in (\overline{H}_0)^c \cap (\overline{H}_1)^c$, we obtain successively:

$$\langle \hat{x}, R[\lambda;T^*|\Xi^*(T^*,\overline{H}_1)]x^*\rangle =$$

$$\langle (\lambda - \overline{T/Y})R(\lambda;\overline{T/Y}|\hat{Z}_0)\hat{x}, R[\lambda;T^*|\Xi^*(T^*,\overline{H}_1)]x^*\rangle = \langle R(\lambda;\overline{T/Y}|\hat{Z}_0)\hat{x},x^*\rangle.$$

Hence g is well-defined. Furthermore, g is analytic on $\mathbb{C}$. For $y \in Y$ and $\Gamma = \{\lambda : |\lambda| = \|T^*|\Xi^*(T^*,\bar{H}_1)\| + 1\}$, we have

$$\langle x,x^*\rangle = \langle x+y,x^*\rangle = \langle\hat{x},x^*\rangle = \frac{1}{2\pi i}\int_\Gamma \langle\hat{x},R[\lambda;T^*|\Xi^*(T^*,\bar{H}_1)]x^*\rangle d\lambda$$
$$= \frac{1}{2\pi i}\int_\Gamma g(\lambda)d\lambda = 0.$$

Consequently, $x \in {}^a\Xi^*(T^*,\bar{H}_1)$ and hence $Z_0 \subset {}^a\Xi^*(T^*,\bar{H}_1) = Y_0$
Similarly, one obtains $Z_1 \subset Y_1$. Thus, by (A.23), X admits a decomposition

$$X = Y_0 + Y_1$$

and in view of (A.24), T has the 1-SDP and hence the SDP, by 6.3. □

B. SOME OPEN PROBLEMS.

In the Notes and Comments on Chapter IV (p.106), a diagram of implications between various types of spectral resolvents is given. Are there more implications? Specifically:

PROBLEM 1. *Is every spectral resolvent of a given operator monotonic? Is every spectral resolvent analytically invariant?*

PROBLEM 2. *Is every monotonic spectral resolvent strongly monotonic? Is every monotonic spectral resolvent analytically invariant? Is every almost localized spectral resolvent a strong spectral resolvent? Is every almost localized spectral resolvent analytically invariant?*

In [W.a] it was shown that there exists a complex Banach space X and an operator $T \in B(X)$, which is decomposable, the adjoint T^* is strongly decomposable but T is not strongly decomposable. This gives rise to

PROBLEM 3. *Does it exist a complex Banach space* X *and a strongly decomposable* $T \in B(X)$, *with* T^* *decomposable but not strongly decomposable?*

PROBLEM 4. *Is it true that* T *is strongly decomposable iff* T^{**} *is strongly decomposable?*

In the study of operators with the SDI, Corollary 17.10 asserts that if a closed operator is endowed with the SDI then it has the SSDP. What about the converse?

PROBLEM 5. *Is there a bounded strongly decomposable operator which is not decomposable with respect to the identity?*

PROBLEM 6. *Is there a closed operator with the* SSDP *which does not have the* SDI?

PROBLEM 7. *Is there an operator which is decomposable with respect to the identity but is not* $\mathfrak{a}$*-scalar for an admissible algebra* $\mathfrak{a}$? (For reference, see [C-Fo.1968]).

BIBLIOGRAPHY

E. Albrecht

[A.1974]* Funktionalkalküle in mehreren Veränderlichen für stetige lineare Operatoren auf Banachräumen. *Manuscripta Math.* 14 (1974), 1-40.

[A.1977] Generalized spectral operators. *Proceedings of the Paderborn Conference on Functional Analysis*, North Holland Math. Studies 27 (1977), 259-277.

[A.1978] On some classes of generalized spectral operators. *Arch. Math.* (Basel) 30 (1978), 297-303.

[A.1978,a]* On two questions of I. Colojoară and C. Foiaş. *Manuscripta Math.* 25 (1978), 1-15.

[A.1979]* On decomposable operators. *Integral Equations Operator Theory* 2 (1979), 1-10.

[A.1979,a] On joint spectra. *Studia Math.* 64 (1979), 263-271.

[A.1981] Spectral decompositions for systems of commuting operators. *Proc. Roy. Irish Acad.* 1981.

[A.1982] Decomposable systems of operators in harmonic analysis. *Toeplitz centennial* (Tel Aviv 1981), *Operator Theory: Advances and Applications*, Vol.4, Birkhäuser Verlag 1982, 19-35.

E. Albrecht and F.-H. Vasilescu

[A-V.1974]* On spectral capacities. *Rev. Roumaine Math. Pures Appl.* 19 (1974), 701-705.

C. Apostol

[Ap.1967] Some properties of spectral maximal spaces and decomposable operators. *Rev. Roumaine Math. Pures Appl.* 12 (1967), 607-610.

[Ap.1968]* Teorie spectrală şi calcul functional. *Stud. Cerc. Mat.* 20 (1968), 635-668.

[Ap.1968,a]* Restrictions and quotiens of decomposable operators in a Banach space. *Rev. Roumaine Math. Pures Appl.* 13 (1968), 147-150.

[Ap.1968,b]* Roots of decomposable operator-valued analytic functions. *Rev. Roumaine Math. Pures Appl.* 13 (1968), 433-438.

[Ap.1968,c]* Spectral decompositions and functional calculus. *Rev. Roumaine Math. Pures Appl.* 13 (1968), 1481-1528.

[Ap.1980]* Functional calculus and invariant subspaces. *J. Operator Theory* 4 (1980), 159-190.

[Ap.1981]* The spectral flavour of Scott Brown's techniques. *J. Operator Theory* 6 (1981), 3-12.

I. Bacalu

[B.1973]* On restrictions and quotients of decomposable operators. *Rev. Roumaine Math. Pures Appl.* 18 (1973), 809-813.
[B.1975]* S-decomposable operators in Banach spaces. *Rev. Roumaine Math. Pures Appl.* 20 (1975), 1101-1107.
[B.1976]* Some properties of decomposable operators. *Rev. Roumaine Math. Pures Appl.* 21 (1976), 177-194.
[B.1976,a] On the restrictions and quotients of S-decomposable operators. *Bull. Math. Soc. Sci. Math. R.S. Roumanie* 20 (68), (1976), 9-13.

W.G. Bade

[Ba.1954]* Unbounded spectral operators. *Pacific J. Math.* 4 (1954), 373-392.

R.G. Bartle

[Bar.1964]* Spectral localization of operators in Banach spaces. *Math. Ann.* 153 (1964), 261-269.
[Bar.1970] Spectral decomposition of operators in Banach spaces. *Proc. London Math. Soc.* 20 (1970), 438-450.
[Bar.1979] Self-adjoint operators and some generalizations. *Operator Theory and Functional Analysis, Research Notes in Mathematics* 38, Pitman Advanced Publishing Program, San Francisco 1979, 36-50.

R.G. Bartle and C.A. Kariotis

[Bar-Ka.1973]* Some localizations of the spectral mapping theorem. *Duke Math. J.* 40 (1973), 651-660.

S.K. Berberian

[Be.1966] Notes on Spectral Theory. Van Nostrand, Princeton 1966.
[Be.1969] An extension of Weyl's theorem to a class of not necessarily normal operators. *Michigan Math. J.* 16 (1969), 273-279.
[Be.1970] Some conditions on an operator implying normality. *Math. Ann.* 184 (1970), 188-192.
[Be.1972] Equivalence of projections. *Proc. Amer. Math. Soc.* 33 (1972), 485-490.
[Be.1974] Lectures in Functional Analysis. Springer-Verlag, New York 1974.

E. Berkson and H.R. Dowson

[Ber-Do.1969] Prespectral operators. *Illinois J. Math.* 13 (1969), 291-315.

[Ber-Do.1971] On uniquely decomposable well-bounded operators. *Proc. London Math. Soc.* (3) 22 (1971), 339-358.

[Ber-Do.1974] On reflexive scalar-type spectral operators. *J. London Math. Soc.* (2) 8 (1974), 652-656.

E. Bishop

[Bi.1957] Spectral theory for operators on a Banach space. *Trans. Amer. Math. Soc.* 86 (1957), 414-445.

[Bi.1959]* A duality theory for an arbitrary operator. *Pacific J. Math.* 9 (1959), 379-397.

A. Brown and C. Pearcy

[Br-Pea.1977]* Introduction to Operator Theory I. Springer-Verlag, New York 1977.

[Br-Pea.1977,a] Jordan loops and decomposition of operators. *Canad. J. Math.* 29 (1977), 1112-1119.

S. Brown

[Bro.1979]* Some invariant subspaces for subnormal operators. *Integral Equations Operator Theory* 1 (1979), 310-333.

I. Colojoară

[C.1968]* Elemente de Teorie Spectrală. Ed. Acad. R.S. România, București 1968.

I. Colojoară and C. Foiaş

[C-Fo.1967]* The Riesz-Dunford functional calculus with decomposable operators. *Rev. Roumaine Math. Pures Appl.* 12 (1967), 627-641.

[C-Fo.1968]* Theory of Generalized Spectral Operators. Gordon & Breach, New York 1968.

R.G. Douglas

[D.1972] Banach Algebra Techniques in Operator Theory. Academic Press, New York 1972.

H.R. Dowson

[Do.1965] Restrictions of spectral operators. *Proc. London Math. Soc.* (3) 15 (1965), 437-457.

[Do.1967] Operators induced on quotient spaces by spectral operators. *J. London Math. Soc.* 42 (1967), 666-671.

[Do.1969]* Restrictions of prespectral operators. *J. London Math. Soc.* (2) 1 (1969), 633-642.

[Do.1978]* Spectral Theory of Linear Operators. *London Math. Soc., Monograph Series* 12, Academic Press, London 1978.

N. Dunford

[Du.1943] Spectral theory I. Convergence to projections. *Trans. Amer. Math. Soc.* 54 (1943), 185-217.

[Du.1952]* Spectral theory II. Resolutions of the identity. *Pacific J. Math.* 2 (1952), 559-614.

[Du.1954]* Spectral operators. *Pacific J. Math.* 4 (1954), 321-354.

[Du.1958]* A survey of the theory of spectral operators. *Bull. Amer. Math. Soc.* 64 (1958), 217-274.

[Du.1966]* A spectral theory for certain operators on a direct sum of Hilbert spaces. *Math. Ann.* 162 (1966), 294-330.

N. Dunford and J.T. Schwartz

[Du-S.1967]* Linear Operators. Part I. Wiley, New York 1967.

[Du-S.1967,a] Linear Operators. Part II. Wiley, New York 1967.

[Du-S.1971]* Linear Operators. Part III. Wiley, New York 1971.

I. Erdelyi

[E.1975]* Unbounded operators with spectral capacities. *J. Math. Anal. Appl.* 52 (1975), 404-414.

[E.1978] The set-spectra of decomposable operators. *Atti Accad. Naz. Lincei Rend. Cl. Sci. Fis. Mat. Natur.* (8) 64 (1978), 32-37.

[E.1978,a] A class of weakly decomposable unbounded operators. *Atti Accad. Naz. Lincei Rend. Cl. Sci. Fis. Mat. Natur.* (8) 64 (1978), 112-117.

[E.1979]* Spectral resolvents. *Operator Theory and Functional Analysis, Research Notes in Mathematics* 38 Pitman Advanced Publishing Program, San Francisco 1979, 51-70.

[E.1980]* Spectral resolutions and decompositions. *Boll. Un. Mat. Ital.* (5) 17-A (1980), 143-148.

[E.1980,a]* Unbounded operators with spectral decomposition properties. *Acta Sci. Math.* (Szeged) 42 (1980), 67-70.

[E.1981]* Monotonic properties of some spectral resolvents. *Libertas Math.* 1 (1981), 117-124.

[E.1982] Spectral decompositions for generalized inversions. *Recent Applications of Generalized Inverses. Research Notes in Mathematics* 66, Pitman Advanced Publishing Program 1982, 261-274.

I. Erdelyi and R. Lange

[E-L.1978]* Operators with spectral decomposition properties. *J. Math. Anal. Appl.* 66 (1978), 1-19.

I. Erdelyi and F.R. Miller

[E-M.1970] Decomposition theorems for partial isometries. *J. Math. Anal. Appl.* 30 (1970), 665-679.

I. Erdelyi and Wang, Shengwang

[E-W.1983]* Spectral decomposition with monotonic spectral resolvents. *Trans. Amer. Math. Soc.* 277 (1983), 851-859.

[E-W.1984]* On strongly decomposable operators. *Pacific J. Math.* 110 (1984), 287-296.

[E-W.1984,a]* A spectral duality theorem for closed operators. *Pacific J. Math.* 114 (1984), 73-93.

[E-W]* Functional calculus and duality for closed operators. *J. Math. Anal. Appl.* (in print).

J. Eschmeier

[Es.1981]* Spektralzerlegungen und Funktionalkalküle für Vertauschende Tupel stetiger und Abgeschlossener Operatoren in Banach Räumen. *Schr. Math. Inst. Univ. Münster* (2) 20, 1981.

[Es.1982] Operator decomposability and weakly continuous representations of locally compact abelian groups. *J. Operator Theory* 7 (1982), 201-208.

[Es.1983] On two notions of the local spectrum for several commuting operators. *Michigan Math. J.* 30 (1983), 245-248.

[Es.1983,a] Equivalence of decomposability and 2-decomposability for several commuting operators. *Math. Ann.* 262 (1983), 305-312.

[Es.1984] Some remarks concerning the duality problem for decomposable systems of commuting operators. *Operator Theory: Advances and Applications*. Proceedings of the eighth conference on Operator Theory, Timişoara/Herculane. Birkhäuser Verlag, Basel 1984.

R. Evans

[Ev.1981]* Embedding C(K) in B(X). *Math. Scand.* 48 (1981), 119-136.

[Ev.1983] Boundedly decomposable operators and the continuous functional calculus. *Rev. Roumaine Math. Pures Appl.* 28 (1983), 465-473.

J.K. Finch

[F.1975]* The single valued extension property on a Banach space. *Pacific J. Math.* 58 (1975), 61-69.

C. Foiaş

[Fo.1960]* Une application des distributions vectorielles à la théorie spectrale. *Bull. Sci. Math.* 84 (1960), 147-158.

[Fo.1963]* Spectral maximal spaces and decomposable operators in Banach spaces. *Arch. Math.* (Basel) 14 (1963), 341-349.

[Fo.1968]* Spectral capacities and decomposable operators. *Rev. Roumaine Math. Pures Appl.* 13 (1968), 1539-1545.

[Fo.1970]* On the maximal spectral spaces of a decomposable operator. *Rev. Roumaine Math. Pures Appl.* 15 (1970), 1599-1606.

[Fo.1972] On the scalar part of a decomposable operator. *Rev. Roumaine Math. Pures Appl.* 17 (1972), 1181-1198.

[Fo.1972,a] Invariant para-closed subspaces. *Indiana Univ. Math. J.* 21 (1972), 887-906.

Ş. Frunză

[Fr.1971]* A duality theorem for decomposable operators. *Rev. Roumaine Math. Pures Appl.* 16 (1971), 1055-1058.

[Fr.1973]* The single-valued extension property for coinduced operators. *Rev. Roumaine Math. Pures Appl.* 18 (1973), 1061-1065.

[Fr.1976]* Spectral decomposition and duality. *Illinois J. Math.* 20 (1976), 314-321.

[Fr.1977]* A new result of duality for spectral decompositions. *Indiana Univ. Math. J.* 26 (1977), 473-482.

[Fr.1983]* A complement to the duality theorem for decomposable operators. *Rev. Roumaine Math. Pures Appl.* 28 (1983), 475-478.

P.R. Halmos

[H.1951]* Introduction to Hilbert Space and the Theory of Spectral Multiplicity. Chelsea, New York 1951.

[H.1967]* A Hilbert Space Problem Book. Van Nostrand, Princeton 1967.

Hou, Xue Zhang

[Ho.1982]* Unbounded operators with decomposable spectrum. *Dongbei Shida Xuebao* (1982), Nr. 4, 11-15.

[Ho.1983]* Unbounded strongly decomposable operators. *Dongbei Shida Xuebao* (1983), Nr. 1, 15-19.

[Ho.1983,a]* Local spectral mapping theorem for closed operators. *Dongbei Shida Xuebao* (1983), Nr. 4, 71-75.

A.A. Jafarian

[J.1977]* Weak and quasi-decomposable operators. *Rev. Roumaine Math. Pures Appl.* 22 (1977), 195-212.

[J.1977,a] Weak contractions of Sz.-Nagy and Foiaş are decomposable. *Rev. Roumaine Math. Pures Appl.* 22 (1977), 489-497.

A.A. Jafarian and F.-H. Vasilescu

[J-V.1974] A characterization of 2-decomposable operators. *Rev. Roumaine Math. Pures Appl.* 19 (1974), 769-771.

S. Kantorovitz

[K.1965]* Classification of operators by means of their operational calculus. *Trans. Amer. Math. Soc.* 115 (1965), 194-224.

[K.1981] Characterization of unbounded spectral operators with spectrum in a half-line. *Comment. Math. Helvetici* 56 (1981), 163-178.

[K.1981,a] Spectrality criteria for unbounded operators with real spectrum. *Math. Ann.* 256 (1981), 19-28.

[K.1983] Spectral Theory of Banach space operators. *Lecture Notes in Mathematics* 1012, Springer-Verlag Berlin, 1983.

T. Kato.

[Kat.1966]* Perturbation Theory for Linear Operators. Springer-Verlag, New York 1966.

S. Kurepa

[Ku.1962]* On n-th roots of normal operators. *Math. Z.* 78 (1962),

285-292.

[Ku.1963]* On operator roots of an analytic function. *Glas. Mat. Ser. II.* 18 (1963), 49-51.

R. Lange

[L.1975] Roots of almost decomposable operators. *J. Math. Anal. Appl.* 49 (1975), 721-724.

[L.1978] Analytically decomposable operators. *Trans. Amer. Math. Soc.* 244 (1978), 225-240.

[L.1979]* Strongly analytic subspaces. *Operator Theory and Functional Analysis, Research Notes in Mathematics* 38, Pitman Advanced Publishing Program, San Francisco 1979, 16-30.

[L.1980] A purely analytic criterion for decomposable operator. *Glasgow Math. J.* 21 (1980), 69-70.

[L.1981]* A generalization of decomposability. *Glasgow Math. J.* 22 (1981), 77-81.

[L.1981,a] Equivalent conditions for decomposable operators. *Proc. Amer. Math. Soc.* 82 (1981), 401-406.

A. Lebow

[Le.1963] On von Neumann's theory of spectral sets. *J. Math. Anal. Appl.* 7 (1963), 64-90.

V.E. Ljance

[Lj.1963] A generalization of the concept of spectral measure. *Mat. Sb.* 61 (103), (1963), 80-120.

[Lj.1963,a] Unbounded operators commuting with the resolution of the identity. *Ukrain. Mat. Z.* 15 (1963), 376-384.

Yu. I. Ljubič and V.I. Macaev

[Lju-M.1960] On the spectral theory of linear operators in Banach spaces. *Dokl. Akad. Nauk SSSR* 131 (1960), 21-23.

[Lju-M.1962] On operators with decomposable spectrum. *Mat. Sb.* 56 (98), (1962), 433-468.

B. Nagy

[N.1977] S-spectral capacities of closed operators. *Studia Sci. Math. Hungar.* 12 (1977), 399-406.

[N.1978]* Operators with the spectral decomposition property are decomposable. *Studia Sci. Math. Hungar.* 13 (1978), 429-432.

[N.1979]* On S-decomposable operators. *J. Operator Theory* 2 (1979), 277-286.

[N.1979-1980]* Closed S-decomposable operators. *Ann. Univ. Sci. Budapest. Eötvös Sect. Math.* 22-23 (1979-1980), 143-149.

[N.1980]* A strong spectral residuum for every closed operator. *Illinois J. Math.* 24 (1980), 173-179.

[N.1980,a] Restrictions, quotients and S-decomposability of operators. *Rev. Roumaine Math. Pures Appl.* 25 (1980), 1085-1090.

[N.1981]* A spectral residuum for each closed operator. *Topics in Modern Operator Theory: Advances and Applications*. Vol. 2. Birkhäuser Verlag 1981, 221-238.

[N.1981,a] Local spectral theory. *Acta Math. Acad. Sci. Hungar.* 37 (1981), 433-443.

[N.1983] Differential operators and spectral decomposition. *Functions Series, Operators*. Vol. I-II (Budapest 1980), 891-917. *Colloq. Math. Soc. János Bolyai* 35, North Holland, Amsterdam 1983.

R.F. Olin and J.E. Thomson

[O-Th.1977] The spectrum of a normal operator and the problem of filling holes. *Indiana Univ. Math. J.* 26 (1977), 541-544.

T.V. Panchapagesan

[P.1967] Unitary operators in Banach spaces. *Pacific J. Math.* 22 (1967), 465-475.

Pei, Yuan Wu

[Pe.1978] Hyperinvariant subspaces of the direct sum of certain contractions. *Indiana Univ. Math. J.* 27 (1978), 267-274.

S. Plafker

[Pl.1970]* On decomposable operators. *Proc. Amer. Math. Soc.* 24 (1970), 215-216.

C.R. Putnam

[Pu.1957]* On square roots of normal operators. *Proc. Amer. Math. Soc.* 8 (1957), 768-769.

M. Radjabalipour

[R.1974]* Growth conditions and decomposable operators. *Canad. J. Math.* 26 (1974), 1372-1379.

[R.1974,a] On decomposition of operators. *Michigan Math. J.* 21 (1974), 265-275.

[R.1975] On decomposability of compact perturbations of operators. *Proc. Amer. Math. Soc.* 53 (1975), 159-164.

[R.1977] Some decomposable subnormal operators. *Rev. Roumaine Math. Pures Appl.* 22 (1977), 341-345.

[R.1978]* Equivalence of decomposable and 2-decomposable operators. *Pacific J. Math.* 77 (1978), 243-247.

[R.1978,a]* Decomposable operators. *Bull. Iranian Math. Soc.* 9 (1978), 1-49.

H. Radjavi and P. Rosenthal

[Ra-Ro.1973] Invariant Subspaces. Springer-Verlag, Berlin 1973.

J. Schwartz

[S.1954]* Perturbations of spectral operators and applications. I. Bounded perturbations. *Pacific J. Math.* 4 (1954), 415-458.

J.E. Scroggs

[Sc.1959]* Invariant subspaces of a normal operator. *Duke Math. J.* 26 (1959), 95-111.

G.W. Shulberg

[Sh.1979]* Spectral resolvents and decomposable operators. *Operator Theory and Functional Analysis, Research Notes in Mathematics* 38, Pitman Advanced Publishing Program, San Francisco 1979, 71-84.

[Sh.1981] Decomposable restrictions and extensions. *J. Math. Anal. Appl.* 83 (1981), 144-158.

R.C. Sine

[Si.1964]* Spectral decomposition of a class of operators. *Pacific J. Math.* 14 (1964), 333-352.

Jon C. Snader

[Sn.1984]* Strongly analytic subspaces and strongly decomposable operators. *Pacific J. Math.* 115 (1984), 193-202.

[Sn]* Bishop's condition (β). *Glasgow Math. J.* (in print).

J.G. Stampfli

[St.1962]* Roots of scalar operators. *Proc. Amer. Math. Soc.* 13 (1962), 796-798.

[St.1966] Analytic extensions and spectral localization. *J. Math. Mech.* 16 (1966), 287-296.

[St.1969] A local spectral theory for operators. *J. Funct. Anal.* 4 (1969), 1-10.

[St.1979]* Recent developments on the invariant subspace problem. *Operator Theory and Functional Analysis, Research Notes in Mathematics* 38, Pitman Advanced Publishing Program, San Francisco 1979, 1-7.

[St.1980]* An extension on Scott Brown's invariant subspace theorem: K-spectral sets. *J. Operator Theory* 3 (1980), 3-21.

Sun, Shan Li

[Su.1982]* Unbounded decomposable operators. *Acta Sci. Nat. Univ. Jilinensis* (1982), Nr.1, 53-65.

[Su.1982,a] Decomposability for a class of non-spectral second order ordinary differential operators. *Acta Sci. Nat. Univ. Jilinensis* (1982), Nr. 4, 9-16.

[Su.1983]* A new proof of equivalence of SDP-operators and decomposable operators. *Acta Sci. Nat. Univ. Jilinensis* (1983), Nr.4, 9-12.

[Su.1984] Decomposability of contractions of class C_{11}. *Acta Sci. Nat. Univ. Jilinensis* (1984), Nr. 4, 53-57.

[Su.1984,a]* A duality theorem for unbounded decomposable operators. *Chinese Ann. Math.* Ser. A, 5 (1984), 49-54.

[Su.1984,b] A decomposable property for weighted shift operators and hyponormal operators on Hilbert spaces. *Chinese Ann. Math.* Ser. A, 5 (1984), 575-584.

Sun, Shan Li and Xu, Feng

[Su-X.1982]* A functional calculus for unbounded decomposable operators. Acta Sci. Nat. Univ. Jilinensis (1982), Nr. 3, 42-50.

Sun, Shan Li and Zou, Cheng Zu

[Su-Zo.1984] Quasi-similarity and the spectrum. *Acta Sci. Nat. Univ. Jilinensis* (1984), Nr. 3, 15-18.

K. Tanahashi

[T.1982] A characterization of decomposable operators. *Tôhoku Math. J.* 34 (1982), 295-300.

[T.1983]* Characterizations of S-decomposable operators on a complex Banach space. *Tôhoku Math. J.* 35 (1983), 261-265.

A.E. Taylor and D.C. Lay

[Ta-La.1980]* Introduction to Functional Analysis. Wiley, New York 1980.

J.L. Taylor

[Tay.1970]* The analytic functional calculus for several commuting operators. *Acta Math.* 125 (1970), 1-38.

[Tay.1970,a]* A joint spectrum of several commuting operators. *J. Funct. Anal.* 6 (1970), 172-191.

F.-H. Vasilescu

[V.1969]* Residually decomposable operators in Banach spaces. *Tôhoku Math. J.* 21 (1969), 509-522.

[V.1971]* Residual properties for closed operators on Fréchet spaces. *Illinois J. Math.* 15 (1971), 377-386.

[V.1971,a]* On the residual decomposability in reflexive spaces. *Rev. Roumaine Math. Pures Appl.* 16 (1971), 1573-1587.

[V.1974] On the decomposability in reflexive spaces. *Rev. Roumaine Math. Pures Appl.* 19 (1974), 1261-1266.

[V.1974,a] Codecomposable operators. *Rev. Roumaine Math. Pures Appl.* 19 (1974), 1055-1059.

[V.1974,b] An application of Taylor's functional calculus. *Rev. Roumaine Math. Pures Appl.* 19 (1974), 1165-1167.

[V.1977] Analytic representations. *Rev. Roumaine Math. Pures Appl.* 22 (1977), 389-401.

[V.1978]* A Martinelli type formula for the analytic functional calculus. *Rev. Roumaine Math. Pures Appl.* 23 (1978), 1587-1605.

[V.1982]* Analytic Functional Calculus and Spectral Decompositions. Ed. Acad. R.S. România, Bucureşti 1982 & Reidel Publishing Company, Dordrecht 1982.

Wang, Shengwang

[W.1979] Operators of $D_{<M_k>}$ type and their resolvents. *Sci. Sinica Math. Issue* (I), (1979), 255-266.

[W.1980] $D_{<M_k>}$ operators and spectral operators. *Chinese Ann. Math.* 1 (1980), 325-334.

[W.1981] The analytic representations of ultradistributions and $D_{<M_k>}$ operators. *Acta Math. Sinica* 24 (1981), 904-912.

[W.1983]* Local resolvents and operators decomposable with respect to the identity. *Acta Math. Sinica* 26 (1983), 153-162.

[W.1983,a] Groups and semigroups of $D_{<M_k>}$ operators on Banach spaces. *Chinese Ann. Math.* 4 A (1983), 505-514.

[W.1984]* On the spectral residuum of closed operators. *Acta Sci. Math.* (Szeged) 47 (1984), 117-129.

[W.1984,a]* On Frunza's paper "A complement to the duality theorem for decomposable operators". *Nanjing Daxue Xuebao (Nanjing University Journal)* (1984), Nr. 1, 20-21.

[W.1984,b] On the spectral mapping theorem of $D_{<M_k>}$ operators. *J. Math. Research and Exposition* 4 (1984), 19-28.

[W]* Theory of spectral decompositions with respect to the identity for closed operators. *Chinese Ann. Math.* (in print).

[W,a]* A characterization of strongly decomposable operators. *Acta Math. Sinica* (in print).

[W,b] On the quasi-decomposability with respect to the identity for closed operators. To be published.

Wang, Shengwang and I. Erdelyi

[W-E.1983]* A duality theorem for unbounded closed operators. *C.R. Math. Rep. Acad. Sci. Canada* 6 (1983), 105-110.

[W-E.1983,a]* Analytically invariant spectral resolvents. *J. Math. Anal. Appl.* 96 (1983), 341-351.

[W-E.1984]* Analytically invariant spectral resolvents of closed operators. *J. Funct. Anal.* 58 (1984), 53-78.

[W-E.1984,a]* Characterizations of closed decomposable operators. *Illinois J. Math.* 28 (1984), 523-530.

[W-E]* On spectral decomposition of closed operators on Banach spaces. To be published.

[W-E,a]* A spectral duality theorem for closed operators III. *J. Sci. Sinica* (in print).

Wang, Shengwang and Guangyu, Liu

[W-Li.1984]* On the duality theorem of bounded S-decomposable operators. *J. Math. Anal. Appl.* 99 (1984), 150-163.

[W-Li] Spectral capacity and decomposability with respect to the identity. *J. Math. Research and Exposition* (in print).

Wang, Shengwang and Sun, Daging

[W-Sun]* Strongly spectral decomposition properties for closed operators. To be published.

Wang, Shengwang; Zou, Cheng Zu and Sun, Shan Li

[W-Zo-Su.1982]* Some properties of decomposable operators. *J. Math. Research and Exposition* 2 (1982), 31-34.

Wang, Shushi

[Wa.1981]* On closed decomposable operators. *J. East China Normal Univ.* 3 (1981), 1-9.

F. Wolf

[Wo.1957] Operators in Banach space which admit a generalized spectral decomposition. *Nederl. Akad. Wetensch. Indag. Math.* 19 (1957), 302-311.

Xu, Feng

[X.1981] A note on spectral decompositions of unbounded operators. *Dongbei Shida Xuebao* (1981), Nr. 2, 29-32.

Xu, Feng and Zou, Cheng Zu

[X-Zo.1982] On Boolean algebra of projection operators. *Dongbei Shida Xuebao* (1982), Nr. 1, 17-22.

[X-Zo.1983] Banach reducibility of decomposable operators. *J. Northeast Normal Univ.* 4 (1983), 61-69.

Zhang, Dianzhou and Wang, Shushi

[Z-Wa.1981]* Unbounded unit-decomposable operators. *J. East China Normal Univ.* 4 (1981), 5-11.

Zou, Cheng Zu

[Zo.1980] Spectral theory of one kind of operators II. *Acta Sci. Nat. Univ. Jilinensis* 1 (1980), 1-8.

On Erdelyi's strong spectral capacities. *Acta Sci. Nat. Univ. Jilinensis* 2 (1982), 8-14

[Zo.1983] Spectral capacity of decomposable operators. *Chinese Ann. Math.* 4 A (1983), 71-78.

[Zo.1983,a] Some properties of quasi-decomposable operators. *Acta Sci. Nat. Univ. Jilinensis* 4 (1983), 13-19.

INDEX

Printed in the United States
By Bookmasters